AF872469

Applied Soil Chemistry

Scrivener Publishing
100 Cummings Center, Suite 541J
Beverly, MA 01915-6106

Publishers at Scrivener
Martin Scrivener (martin@scrivenerpublishing.com)
Phillip Carmical (pcarmical@scrivenerpublishing.com)

Applied Soil Chemistry

Edited by

Inamuddin

Department of Applied Chemistry, Aligarh Muslim University, Aligarh, India

Mohd Imran Ahamed

Department of Chemistry, Aligarh Muslim University, Aligarh, India

Rajender Boddula

National Center for Nanoscience and Technology, Beijing, China

and

Tariq Altalhi

Department of Chemistry, College of Science, Taif University, Taif, Saudi Arabia

Scrivener Publishing

WILEY

This edition first published 2021 by John Wiley & Sons, Inc., 111 River Street, Hoboken, NJ 07030, USA and Scrivener Publishing LLC, 100 Cummings Center, Suite 541J, Beverly, MA 01915, USA
© 2021 Scrivener Publishing LLC
For more information about Scrivener publications please visit www.scrivenerpublishing.com.

All rights reserved. No part of this publication may be reproduced, stored in a retrieval system, or transmitted, in any form or by any means, electronic, mechanical, photocopying, recording, or otherwise, except as permitted by law. Advice on how to obtain permission to reuse material from this title is available at http://www.wiley.com/go/permissions.

Wiley Global Headquarters
111 River Street, Hoboken, NJ 07030, USA

For details of our global editorial offices, customer services, and more information about Wiley products visit us at www.wiley.com.

Limit of Liability/Disclaimer of Warranty
While the publisher and authors have used their best efforts in preparing this work, they make no representations or warranties with respect to the accuracy or completeness of the contents of this work and specifically disclaim all warranties, including without limitation any implied warranties of merchantability or fitness for a particular purpose. No warranty may be created or extended by sales representatives, written sales materials, or promotional statements for this work. The fact that an organization, website, or product is referred to in this work as a citation and/or potential source of further information does not mean that the publisher and authors endorse the information or services the organization, website, or product may provide or recommendations it may make. This work is sold with the understanding that the publisher is not engaged in rendering professional services. The advice and strategies contained herein may not be suitable for your situation. You should consult with a specialist where appropriate. Neither the publisher nor authors shall be liable for any loss of profit or any other commercial damages, including but not limited to special, incidental, consequential, or other damages. Further, readers should be aware that websites listed in this work may have changed or disappeared between when this work was written and when it is read.

Library of Congress Cataloging-in-Publication Data

ISBN 978-1-119-71018-9

Cover image: Pixabay.Com
Cover design by Russell Richardson

Set in size of 11pt and Minion Pro by Manila Typesetting Company, Makati, Philippines

Printed in the USA

Contents

Preface

Soil chemistry refers to the chemical reactions in soils that affect the growth and nutrition of plants. Applied soil chemistry is an interdisciplinary field covering soil, water, plants and atmosphere, which impacts plant, animal and human health. Water and nutrients are provided in different types of soil that are home to microorganisms and many other creatures and plants. The properties of these soils affect the crop production of agricultural fields; therefore, this discipline provides support to the sustainable agricultural management of soils.

State-of-the-art information regarding applied soil sciences is explored in this book. In addition to the fundamentals of soil chemistry, model concepts, principles, chemical reactions, functions, chemical recycling, chemical weathering, acid-base chemistry, carbon sequestration, and nutrient availability of soils are highlighted. Also included among the topics are the chemistry of heavy metals in soil environments, ion-exchange processes on clay, along with relevant analytical tools and applications. This book will help the reader understand soil characteristics by targeting soil chemical reactions and interactions and their applications. Since the chapters were written by noted professionals in the field, it will be an excellent reference guide for students, faculty, researchers and professionals in the field of environmental science, earth science, soil chemistry, and agroecology. The subject matter covered in each chapter is summarized below.

Chapter 1 provides details on the significance of soils as a carbon store. Retaining and ideally boosting carbon in soils helps to inhibit its buildup in the atmosphere. It takes some time for soil carbon to become mineralized, whereas soil erosion and tillage continuously release some soil carbon into the atmosphere.

Chapter 2 provides background information on the chemical weathering of minerals. A discussion of the weathering sequence of minerals from the soil mainly sheds light on the factors which control the rate of chemical weathering, including temperature and time factors, biotic process, oxidation, reduction, comparative stability of various minerals, water, leaching, and acidity.

Chapter 3 discusses the effect that agricultural management systems have on soil health by evaluating reduced processing systems, organic fertilizers, and biological control of weeds, along with their effect on the bioeconomy, including agriculture, forestry, fishing, aquaculture, woodworking, biorefineries, and nanobiotechnology. The data provided is based on the evaluation of indicators of physical, chemical, biological, microbiological, and biochemical soil health. The benefits of soil resilience and better adaptation to extreme events are also included.

Chapter 4 discusses the advances in soil chemistry which are based on the analysis of the soil at an *in-situ* level and the presence of carbon speciation in soil. The analysis of the soil using sensors, the internet of things, machine learning, artificial intelligence, and drones with big data are also discussed.

Chapter 5 details the various components of soil, including solid, liquid and gaseous phases. Different soil characteristics like structure, soil color, texture, bulk density and particle size are discussed in detail. These are important parameters for understanding soil behavior. Sorption behavior of soil for removal of heavy metals is also discussed.

Chapter 6 relates to inclusive edaphology and environmental reactions. Parameters such as composition, textures, soil organic matter, salinity, acidity and chemical reaction of soil are discussed. The main focus is on different kinds of contaminants like pesticides, modern agriculture, synthetic chemicals and their effects on plant growth.

Chapter 7 describes fertilization and fertilizer types. The purpose of fertilization and its application methods are discussed. Fertilizers can be classified into two types depending on the source from which they are obtained: organic (natural) fertilizer and chemical fertilizer. The importance of the use of fertilizers in agricultural lands for sustainable agriculture is explained.

Chapter 8 discusses several approaches which are used to control the overabundance of heavy metals present in soil. The focus of this chapter are the techniques utilized in the past few years to estimate heavy metal content and its mitigation process.

Chapter 9 details the importance and benefits of modeling studies about the retention and mobilization of pollutants. Recently applied models and primary outcomes of modeling studies are also discussed. Additionally, several personal computer programs, which are used for running models, are exemplified in this study.

Chapter 10 provides details of the different laws of soil chemistry. The reaction of ions with the variable charged mineral surfaces, organic matters and clay minerals are discussed. The role of water movement through saturated and unsaturated media is also discussed, with a major focus on gravitational and capillary forces.

Chapter 11 discusses various aspects of assessing the quality of soil using mechanical and physicochemical parameters. Mechanical parameters, which play an essential role in plant growth, are also discussed. There is also a discussion of ion-exchange properties, which is the primary focus of the studies on physicochemical parameters.

Chapter 12 explains the necessary soil functions that could lead to an increase in agricultural production. Essential soil biological processes and their relationship with soil pH are also discussed along with microbial ecophysiological indicators and activities of soil enzymes.

Chapter 13 provides detailed information on the application of pesticides in agriculture. It also elaborates on the numerous types of pesticides, including neonicotinoids, pyrethroids, organochlorines, organophosphates, triazines, carbamates, and pyrethroids. The merits and demerits that are associated with chemical pesticides are also highlighted. This chapter also includes the modes of action and the detrimental effects of these pesticides.

Inamuddin, Mohd Imran Ahamed,
Rajender Boddula and Tariq Altalhi
February 15, 2021

1

Potential and Challenges of Carbon Sequestration in Soils

Erfan Sadatshojaei[1]*, David A. Wood[2] and Mohammad Reza Rahimpour[3]

[1]Department of Chemical Engineering, Shiraz University, Shiraz, Iran
[2]DWA Energy Limited, Bassingham, Lincoln, United Kingdom
[3]Department of Chemical Engineering, Shiraz University, Shiraz, Iran

Abstract

Terrestrial soils, by volume, represent the most significant land-based carbon store on our planet. Over time, soils absorb carbon from a wide range of organisms as they respire during life and decompose after their demise. Carbon currently residing in the upper soil layers constitutes more than the combined quantity of carbon in land-surface vegetation and the atmosphere. Retaining and ideally boosting that carbon store in soils and preventing that carbon entering the atmosphere is of paramount importance in the fight against climate change. Almost 50% of global soils within about 1 m of the surface have been disturbed by agriculture releasing at least some of the carbon they store to the atmosphere. Carbon ideally needs to become mineralized in soils if it is to be stabilized and sequestered in the subsurface over the long term. Unfortunately, a significant portion of carbon in soils has a relative rapid turnover time, or low residence time, and is returned to the atmosphere as carbon dioxide via soil respiration processes. Whereas, it takes much longer for some of the soil carbon to be converted to stable mineralized forms. Soil erosion, as well as tillage, plays a significant role in releasing some soil carbon to the atmosphere. Converting significant areas of croplands and grazing lands to forests, grassland, and wetlands is the best option currently available for increasing the soils uptake of carbon from the atmosphere. Additionally, plant large quantities of perennial deep-rooted, fast growing bioenergy crops, such as switchgrass and miscanthus, can increment the carbon storage potential of grassland soils. The aggressive implementation of such actions has the potential to increase global soil carbon storage by between 0.5 and 2.0 Pg C a^{-1} for several decades. This could

**Corresponding author*: erfan.sadatshojaei@gmail.com

Inamuddin, Mohd Imran Ahamed, Rajender Boddula and Tariq Altalhi (eds.) Applied Soil Chemistry, (1–22) © 2021 Scrivener Publishing LLC

only be achieved in association with large-scale reforestation and robust steps to mitigate anthropogenic soil disturbance and natural erosion.

Keywords: Soil as a carbon store, carbon turnover time, climate change mitigation, soil disturbance, organic matter decomposition, humification and peat formation, carbon mineralization, soil C sequestration potential

List of Abbreviations and Units

1 petagram (Pg) = 10^{15} grams = 1 gigatonne
1 kilogram (kg) of carbon is equivalent to 44/12 of carbon dioxide
1 hectare (ha) = 107,639 square feet = 10,000 square meters
1 acre = 43,560 square feet = 4047 square meter
10,000 parts per million (ppm) = 1%
CO_2 = carbon dioxide
NPP = net primary production

1.1 Introduction

Today, more than ever, the impact of the advancements in a wide range of technologies and artificial intelligence are influencing progresses and development of human life across the world. In this regard, we can point out ultrasonic application [1], carbon dioxide issue [2], medical research [3–8], and new chemical methods [9] are all advancing rapidly with their impacts being felt more widely. "Sequestration" of combines both the capture of carbon dioxide (CO_2) and its long-term isolation from the atmosphere and ocean, storing it safely and securely for thousands of years.

Carbon sequestration in soils, to absorb some of the unwanted CO_2 in the atmosphere, mainly involves adopting improvements in land management. This means adopting practices, on a large scale, that convert more atmospheric CO_2 into carbon stored in soils than current practices achieve. The main potential to improve carbon management techniques applies to cropland and grazing lands [10]. These improved land use and carbon management techniques strive to increase the rate of biomass entering the soils and/or by reducing the rates of turnover of organic carbon already residing in the soils and by increasing the quantity of soil carbon that becomes mineralized. Through carbon sequestration in soils, CO_2 is to a degree stabilized in soils on a semi-permanent basis. However, to achieve this, the CO_2 needs to be converted into other materials. These chemical changes are initiated primarily through organic processes. CO_2 is involved in a complex cycle which includes a circulation through the soil influenced by a range of micro-biological activity. Through this circulation, CO_2 becomes available in soils to be dissolved by percolating rainwater. This leads to the formation of carbonic acid in near surface fluids.

On a global scale, soils store about 1,500 Pg (petagrams, equivalent to 1,500 gigatonnes) of organic carbon to a depth of one meter, increasing to 2,400 Pg to a depth of 2 m [11]. This means that carbon residing in upper soil layers amounts to more than the combined quantity of carbon in land-surface vegetation and the atmosphere. A little less than 50% of soils globally have been or are in use for agriculture, both cropland and grazing land. The soils involved in cropland activity have almost all been disturbed by some form of tillage. Organic matter within soils can vary between about 1% and 10%.

Subsequently, that carbonic acid reacts with basic cations leading to the creation of secondary carbonates in the short-term, on the scale of years, forming mineralization in near-surface rocks, leading to sustained processes that persists over geological timescales. The creation of secondary carbonates comes mainly from sub-surface weathering and diagenesis reactions with silicate minerals containing calcium and magnesium. Such reactions generate free positively charged ions (cations). Many of these free cations go on to combine with CO_2 to form carbonate minerals, particularly calcite and dolomite [12]. However, these pervasive carbonate forming diagenetic processes tend to progress too slowly in their natural cycles to be practically exploited for carbon sequestration purposes. Nevertheless, they do involve substantial quantities of CO_2, particularly in alkaline and saline soils present in dry and semi-dry zones [13]. Consequently, the inorganic sub-surface carbon cycle cannot be considered as significant or viable for rapid carbon sequestration in the soils typically found in the soils of wet and temperate zones.

On the other hand, organic carbon can cycle through soils, some returning to the atmosphere very rapidly. In the organic dimensions of the carbon cycle, atmospheric carbon dioxide is stabilized through the photosynthesis conducted by plants, algae, and cyanobacteria to form a range of organic compounds. Although the living organisms initially form glucose during photosynthesis, they transform it into diverse organic compounds, such as cellulose, hemicellulose, and lignin mostly; materials that are useful for biological growth and tissue formation. However, other complex organic materials such as protein, lipids, including more intricate compounds used to provide various benefit to plants and bacteria, are also formed. Land plants direct a significant portion of photosynthetic products to their

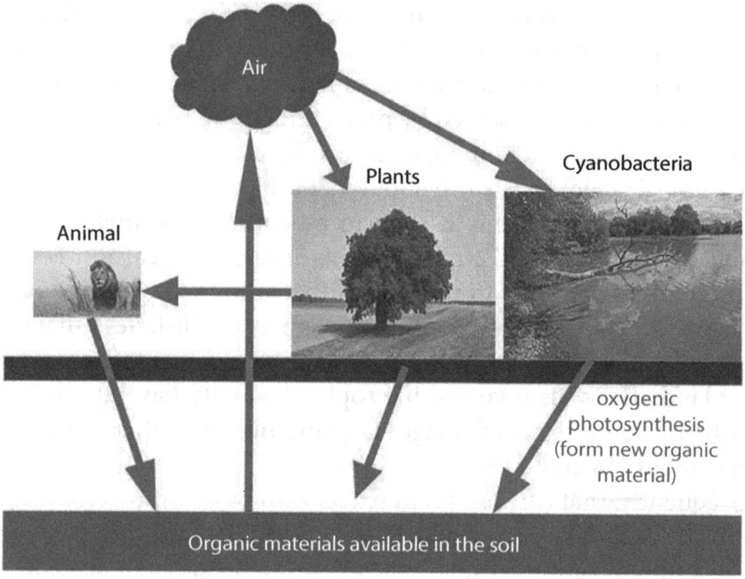

Figure 1.1 Schematic diagram illustrating the biological contributions to the carbon cycle via terrestrial soils.

roots, some of which are released to the soil as soluble carbon compounds; products termed as rhizoexudates [14].

When plants and bacteria die, their organic constituents are dispersed in soils through decomposition by soil micro-organisms. That decomposition releases much of the CO_2 they captured during photosynthesis making its way out of the soil to return to the atmosphere (Figure 1.1). This organic matter-soil decomposition cycle contributes CO_2 output to overall soil respiration that includes respiration of plant root and flora and fauna that live in the soil. In addition to the contributions of plants, algae, and cyanobacteria to the carbon cycle through soils, there is a substantial sub-cycle that is related to contributions from animals. The animals consume CO_2 in the form of food, with animal excrements and corpses returning to the soil and being decomposed along with plant, algae, and cyanobacteria remnants.

1.1.1 Soil Decomposition Processes

Animals that live in the soil vary from clearly visible spineless animals such as woodlice, centipedes, and earthworms, to smaller, microscopic-scale animals, the mesofauna, including mites (arthropods), springtails

(Collembola a hexapod), and enchytraeidae worm-like (microdrile oligochaete) creatures. Some of the smallest animals in the soil, such as nematodes and protozoa, are among the most effective in the soil decomposition processes [15]. Soil organisms of all sizes and types collectively consume plant, animal, and bacterial debris in the soil. They do this by communition, or the reduction of material from one average particle size to a smaller average particle size, using various physical and biochemical techniques. Fungi and bacteria play a key role in breaking down the structural fabrics of plant materials. These groups are able to convert cellulose and lignin into soluble materials applying complex enzymes to do so. Subsequently, the soluble materials produced are absorbed by the organisms and further metabolized.

Initially, dead plant material located above the ground are, for the most part, decomposed above ground on the soil surface. The soil organisms, weather, and industrial-scale anthropogenic mechanical process such as ploughing, play the substantial role in initiating above-ground and near-surface soil decomposition. In some specific cases, for instance, peat formed in bogs and swamps, the dead plant substances stay at the surface of the soil without time to progress through the complete decomposition process. Instead, it becomes rapidly inhumed by other dead plant substances being added from above, isolating it from abundant oxygen supplies.

A consequence, at completion of the soil decomposition processes, is that carbon is ultimately conveyed from the decaying matter into fungi and soil bacteria. This material is known as microbial biomass. Microbes generate and use this biomass to provide their energy requirements and to create new microbial biomass for growth. That carbon used for energy is converted to CO_2 and contribute to soil respiration. However, that portion of the carbon transformed into new microbial biomass, ultimately, is itself consumed or decays upon the demise of the micro-organism and contributes to the ongoing cycle of decomposition. Each successive step in the soil decomposition process involves the consumption of dead biomass by soil organisms, mainly fungi and soil bacteria. Thus, specific carbon molecules pass through many cycles of decay and ultimately end up, over time, either re-emitted to the atmosphere (soil respiration) or fixed by carbon mineralization in the subsurface. On a global scale, the amount of carbon mineralized by soil decomposition is approximately equal to the carbon arriving the soil less the amount re-emitted by soil respiration. Soil respiration returns about 60 Pg a^{-1} (petagram per year; equivalent to 60 gigatonnes) of carbon to the atmosphere, which is about half of the carbon entering the soil [14].

Total carbon accumulating in soils, termed as net primary production (NPP) from organic sources, varies significantly depending on local climatic conditions, vegetation zones, and ecosystems. NPP varies from about 0.5 t C $ha^{-1} a^{-1}$ (tonnes of carbon per hectare per year) in deserts to about 4 t C $ha^{-1} a^{-1}$ in grasslands to about 10 t C $ha^{-1} a^{-1}$ in tropical rainforests [14]. A direct positive relationship also exists between NPP and the magnitude of carbon released by soil respiration [16].

1.1.2 Organic Compounds Present in Soils

Organic organisms and compounds present in soils are diverse, and the provenance of some of the organic molecules, not present in organisms or identifiable fragments of organisms, is often uncertain. The type and quantity of organic material accumulating in soils is influenced by seasonal weather and biological life cycles. Decomposition processes tend to target the simple biochemical molecules initially; amino acids, nucleic acids, proteins, and sugars. Degradation of the more complex structured molecules, cellulose, hemicellulose, pectins, and polymers takes much longer. Lignins, present in the cell walls of wood and tree bark, takes the longest time to be broken down.

Most biomass consists of complex mixtures of the organic materials mentioned all decaying at different rates and forming physical and chemical mixtures. The presence of lignin in such complex mixtures tends to slow down the decomposition of the components that on their own would decay more rapidly. The degradations of mixtures of organic materials containing cutins and tannins can also be retarded in a similar manner.

Microbes constitute up to about 3% of the organic matter in soils. Bacteria and fungi have their own complex chemical makeups that distinguish them from plant and animal derived biomass. Melanins, derived from some fungi, are not degraded easily and are partly responsible for the dark coloration of some soils. Microbial actions tend to remove visual signs of the structures of organisms they decay turning the organic material into a biochemical soup revealing only their microbial provenance. Some non-biological synthesis of chemical bonds does occur in this biochemical soup, in the presence of oxygen and groundwater of varying acidity, contributing to the humification of soils. The humic materials generated to be stable and not susceptible to fragmentation by biologically produced enzymes [17]. During decomposition the carbon quantity of the organic materials in soils increases from about 40% (undecayed plant material) to about 60% (fully humified soil). Humified soil is not just organic material, it involves mixtures with various inorganic minerals from the weathered outcropping rocks on which it resides, particularly the finer grained clay minerals. The combination of decayed organic material with inorganic minerals, physically as grain coating and chemically as mineralization reactions, aids the structural integrity of soils and often helps to inhibit mechanical processes involved in soil erosion.

1.1.3 Cycle Time of Carbon in Soils

Despite a global balance between carbon inflow from biological sources to a soil and CO_2 output to the atmosphere through soil respiration, that process tends not to be a smooth sequential progression when observed locally in specific soils. On the contrary, carbon introduced in one seasonal cycle may take multiple seasonal and/or annual cycles before it is decayed step-by-step and ultimately becomes mineralized. The CO_2 soil respiration output at any point in time comes cumulatively from organic material, at various stages of decay, introduced to the soil during multiple seasonal and annual cycles. The turnover time for carbon in soils varies significantly depending on its stage of maturity and can be determined precisely by carbon isotope dating (^{14}C) techniques. For immature soils, rich in recently demised and introduced biological material, the turnover time is no more than 5 years and could be less than one year. For mineralized soil mixed with inorganic minerals, the turnover time is likely to amount to several decades. The most mature humified soils, from which there is very little carbon inflow and outflow, the carbon turnover times can be measured in thousands of years, meaning that they represent almost inert systems from the perspective of carbon flow [18].

It is often useful to establish turnover times for specific soils. In practice, individual soils will contain components and/or layers displaying a wide range of carbon turnover times. The turnover time of a particular soil layer tends to be influenced by the vegetative geographic zone which determines its sustained humidity and temperature and nutrient content. However, soil instability, due to impacts of extreme seasonal climatic swings and/or severe weather events that can frequently disturb soils, for example, by increasing leaching rates by ground water, do substantially influence carbon turnover times in some cases.

He *et al.* (2016) [19] presented ^{14}C dating measurements on 157 globally distributed soil profiles sampled to 1-m depth. Their results revealed that most existing Earth system models tended to substantially underestimate the mean age of carbon in all common soil types. Moreover, that underestimate was more than six-fold (430 ± 50 years versus 3,100 ± 1,800 years). This finding is not good news for soils as a potential carbon sequestration store, implying that models were making two-fold overestimates of soils global carbon sequestration capabilities. The results indicate that carbon stabilization is a slow process in many soils with long turnover times. This makes soils relative slow-to-develop and passive reservoirs, except on geological time scales, meaning that they cannot absorb large quantities of carbon over short periods of time. This limits soils potential to be a major part of a short-term carbon sequestration solution aimed at removing carbon from the atmosphere on a timescale of decades and over the duration of coming century, which is required to rapidly reverse rising carbon levels in the atmosphere.

The UN's Intergovernmental Panel on Climate Change (van Diemen, 2019) [20], among other bodies (e.g., Halldorsson *et al.*, 2015 [21]), had previously developed its models for carbon sequestration by soils assuming substantially shorter carbon cycling times. This erroneous assumption led the IPCC to suggest if deforestation on a global scale could be halted about 40 ppm of CO_2, about 10% of current levels could be sequestered into soils from the atmosphere. It was believed that combined with major global changes in agricultural practices even more carbon could be absorbed by soils. There is now a more realistic recognition that the carbon absorption ability soils and their turnover periods for carbon cannot be substantially increased in the short-term from the prevailing slow rates [19].

1.2 Influences Impacting Carbon Stabilization Rates in Soils

The amount of organic material present in soils is influenced by many factors (Figure 1.2). Establishing knowledge and understanding of the key influencing factors are essential in determining how carbon uptake and retention of certain soil types can be improved from a long-term carbon capture and sequestration perspective. In a specific soil, the alteration rate of the unit quantity of organic material it contains is established by subtracting the degradation rate from the input rate of biologically derived material. The organic matter degradation rate is positively correlated with the quantity of organic matter present in the soil. This relationship means that progressively the degradation rate converges to the input rate as a soil matures to establish an environmentally adjusting equilibrium with the organic matter remaining constant from that point. The quantity of carbon resident and fixed in a unit mass of soil can therefore be enhanced by either reducing the rate of its degradation of organic matter or accelerating the input of organic matter. On the other hand, more carbon can also be fixed

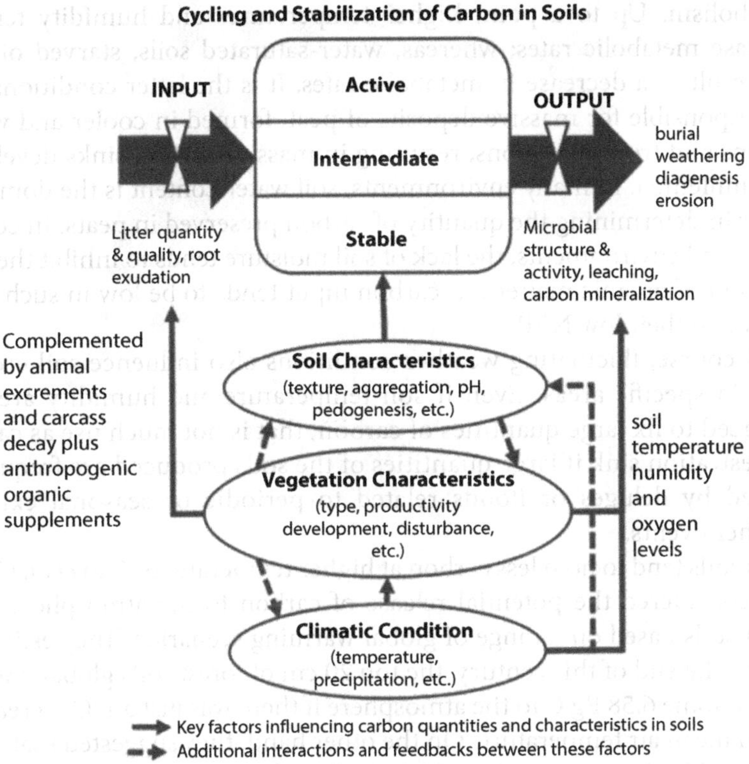

Figure 1.2 Schematic diagram showing how the key variables of climate, vegetation, and soil characteristics impact organic matter concentrations in soils (modified after Sun *et al.*, 2019 [22], who developed the diagram with a specific focus on forest ecosystems).

in a soil system if more mature humified material is continuously produced. This is because the humified material tends to be inert with carbon contributing to its formation, but unlikely to leave it for millennia until it is eroded by geological processes or excavated by anthropogenic activities.

1.2.1 Weather Conditions and Fluctuations

Prevailing weather conditions substantially influence the quantity of carbon present in soils. This is not simply due to differences in specific climatic zones and geographical environments that clearly plays a substantial role. It is well established that the carbon content of soils reduces from the equator to the poles, mainly as a consequence of climatic variations [23]. Specifically, it is biophysical variables such as humidity and temperature conditions in a soil that determine the rates of microbial and bacterial metabolism. Up to a point higher temperatures and humidity tend to increase metabolic rates; whereas, water-saturated soils, starved of oxygen result in a decrease in metabolic rates. It is the latter conditions that are responsible for massive deposits of peat, formed in cooler and wetter regions and tropical regions, resulting in massive carbon sinks developed over millennia. In many environments, soil water content is the dominant factor in determining the quantity of carbon preserved in peats. In cool or warm arid environments, the lack of soil moisture tends to inhibit the degradation of organic matter but carbon input tends to be low in such soils, leading to their low NPP.

Of course, fluctuating weather conditions also influence soil erosion rates in specific areas. Even if soil temperature and humidity are well balanced to fix large quantities of carbon, that is not much use as carbon sequestration sink if large quantities of the soils produced are frequently eroded by deluges or floods related to periodic or seasonal extreme weather events.

As soils tend to hold less carbon at higher temperatures, Sun *et al.* (2019) [22] considered the potential release of carbon to the atmosphere from forest soils based on a range of global warming scenarios. They estimated that by the end of this century, the top 20 cm of forest soils globally would release some 6.58 Pg C to the atmosphere if there was just a 1°C increase in global mean air temperature. On the other hand, they suggested that those soils could release as much as 26.3 Pg C to the atmosphere if there was a 4°C increase in global mean air temperature. In either case, the addition of carbon to the atmosphere would likely exceed carbon uptake from reforestation and forest growth.

Rate of decomposition and availability of carbon in soil to microorganisms is often used to distinguish two type of organic carbon in soil: labile and recalcitrant [24]. Labile soil carbon tends to be associated with microbial biomass, dissolved organic matter that is easily oxidized and broken down. Recalcitrant organic carbon in the soil are compound that are resistant to microbial decomposition. Recalcitrant organic carbon tends to be associated with soil mineral particles. Zhang and Zhou (2018) [25] estimate that more than 80% of mineralized soil carbon is derived from the recalcitrant pool of soil organic matter in the temperate forests of northern China. In those forests, the quantity of mineralized soil organic carbon slightly increased with soil moisture content. Mineralization of soil carbon clearly plays a key role in the carbon cycle in terms of determining soil's ability to store quantities of carbon in a stable manner over long periods of time. That ability is strongly influenced by temperature and moisture content.

1.2.2 Plant and Natural Biomass Inputs

The quantity of carbon in a soil is positively correlated with its input rate of biologically derived material. Increasing that input rate is therefore likely to increase a soil carbon stock. In agricultural regions, this can be achieved, to an extent, by minimizing the time the land remains fallow without vegetation, although in arid regions fallow periods are often essential to restrict overuse of limited water resources. Introducing higher-yield, fast-growing crops and grass leys into crop rotation sequences tends to increase the amount of organic material entering the soil due to the higher biomass production above ground [26]. For instance, a crop rotation including deep-rooted grasses enhanced the organic content in soils from savannah environments by as much as 70 t C ha^{-1} [27]. Ensuring that maximum quantities of the residual wastes from crop harvests are allowed back into the soil in agricultural regions enhances soil organic content [28]. Increasing biomass production with the aid of nitrogen-rich fertilizers tends to work well in temperate climates but not so well in tropical climates [29]. However, excess nitrogen in rivers and in the oceans surrounding major river deltas is known to have a detrimental effect on biodiversity.

1.2.3 Organic Enrichment Treatments

Introducing organic supplements to agricultural land that is over and above the carbon input resulting from the rotation cycle of crops grown on the land does act to increase the resulting carbon stored in a soil. The simplest way to achieve this is to extensively mulch land with composts and manures. In some developing countries and communities, manure/dung is dried and combusted as low-grade domestic and communal fuel, rather returned to the land. Composted municipal waste and sewage sludges are also applied as organic supplements to farmland at agricultural scales. These can introduce harmful chemicals, such as heavy metals, into soils, so their compositions have to be carefully monitored. Such treatments can also have negative impacts on the biodiversity of both flora and fauna supported by the land if applied inappropriately. Arable farmland generally responds well to such supplements, but grasslands growing on poor soils can be damaged by them [30, 31].

1.2.4 Tilled and Ploughed Agricultural Land

Whereas tilling prepares land for cultivation by mixing particles as a till is dragged through a soil ploughing tends to fragment clumps of soil and bury weeds and crop residues by penetrating deeper into the soil layers. Both activities are generally referred to as tillage. These activities typically result in higher carbon losses from cultivated soils than undisturbed grassland or forest soils, which typically possess the lower levels of carbon in their undisturbed condition. Deforestation of land to create new land for cultivation often leads to their soils being rapidly degraded in carbon and certain mineral components by groundwater leaching and enhanced microbial metabolism that increases the rate of soil respiration. Increased soil oxygen levels and changes in their temperature and humidity by tillage disruption are typically the causes of enhanced carbon losses from soils, both in the form of CO_2 to the atmosphere and as leached organic compounds deeper into the subsurface [32]. Replacing ploughing with less disruptive tilling using shallow-tine equipment can reduce carbon losses from a soil substantially. Whereas ploughing can result in greater than 25% carbon loss, replacing it with shallow tilling can result in just 5% or 6% carbon loss [33]. Soil carbon content can be reduced rapidly the introduction of arable cultivation techniques involving some level of tillage. Buhre *et al.* (2005) [34] recorded carbon losses of up to 11% from soils subjected to just one cultivation cycle.

1.2.5 Pasture Managed for Livestock Grazing

Grasslands are typically able to sustain higher soil carbon contents than soils of arable land not treated with organic supplements. Moreover, best practices in grazing land management can improve the organic density of grass sward, that is, the surface soil layer constituted by a mass of grasses and their roots. This, in turn, may increase soil carbon levels by greater than 1.0 t C ha^{-1} a^{-1} [13, 28]. However, poor grassland management, which tends to be the norm in deforested areas, associated with over-grazing with livestock exacerbates carbon loss, soil erosion, and reductions in biodiversity at the macro and micro levels.

1.2.6 Irrigated Arable Lands and Their Associated Drainage

Water movements through soils and water retention by soils have complex interactions with soil carbon levels. Whereas, decomposition rate of organic matter in soils tends to be reduced by waterlogging, the introduction of artificial drainage, for example, draining peat bogs and fens, often rapidly degrades their carbon contents and microbial activity. Such actions led to the thickness of peat in eastern England reducing by some 4 m due to shrinkage due to water reduction in the peat and by reduced carbon content with more CO_2 lost to the atmosphere [35]. Remedial actions involving artificially increasing water table levels is able can arrest or slow down carbon losses from artificially drained soils [36]. On the other hand, for the most part, projects to artificially irrigate land have positive outcomes for soil carbon contents. They achieve this by raising crop yields and increasing organic matter from crop residues and/or organic supplements making its way into the soil [28, 37].

1.2.7 Uncertain Impacts of Soil Erosion and Redistribution on Its Carbon Store

Soils are not static systems. Climatic and geological processes act continuously to impact soils. Several of these processes disturb and redistribute soils due to erosion and transportation actions associated with the forces exerted by blowing winds and/or flowing water. Soil erosion ultimately moves the organic content of soil from its place of original formation to another, or ultimately distribute portions of soil from one location to several subsequent locations. Similar to the anthropogenic consequences of tillage, natural erosion processes lead to considerable soil disruption that is likely to increase the degradation rate of its organic matter [38]. However, if soils are merely displaced by a single erosion event, for example, by a landslip, relatively small amounts of carbon loss are likely to result. Unfortunately, the continuous and repeated nature of erosive forces and events leads soil components to gradually make their way, via fluvial actions, into ocean sediments. Although, the carbon that does ultimately arrive in ocean sediments may then be buried for many millennia and isolated from the ocean and atmosphere [38], much of the carbon temporarily fixed by the original soil formation processes is lost or partially degraded along the way. This makes it almost impossible to quantify on a global scale what quantities of carbon stored in soils is ultimately returned to the atmosphere by the geological forces of erosion. In the short term also, the quantities of crop residues entering a soil system is reduced by the actions of erosion at the surface, typically resulting from extreme or seasonal weather-related disturbances.

1.2.8 Fire Impacts on Soil Characteristics

Across the world, fire is relatively frequent but intermittent natural phenomenon impacting large tracts of land. In addition to natural causes of land fires (lightning strikes, volcanic activity, etc.) there are anthropogenic causes such as crop residue burning, forest clearance, accidents, and arson.

Isolated, high-latitude wetlands are much less likely to experience natural land fires caused by lightning strikes [39] than hotter drier regions close to human population centers, where substantial quantities of dry biomass at the land surface is more easily ignited.

Seasonal burning to contain vegetation is a land management technique used in some areas. Although this removes biomass and temporarily reduces carbon input to the underlying soils [40], this practice can ultimately stimulate new, more rapid vegetation growth that will, for a time, enhance carbon input to the soil. The addition of some "charcoal" or charred carbon-rich material to the soil following episodes of burning also contributes to the inert carbon matter in the soil. This component, and artificially produced biochar additions, acts to stabilize some of the carbon within fire-affected or biochar-treated soils. Whether induced naturally or for land management purposes, there is a risk that land fires can ultimately ignite the underlying soils and peats. If this occurs, it releases huge quantities of CO_2 to the atmosphere, and it reduces or removes the carbon store in the soils that have taken millennia to accumulate. Such outcomes have been observed associated with forest clearances in Brazil and Indonesia and are likely intermittent natural events over geological time scales.

1.3 Carbon-Sequestration Potential of Specific Vegetation Zones and Ecosystems

Fundamentally, each type of land-surface vegetation type has characteristic carbon quantities and fabrics in its underlying soils. For instance, in temperate climates carbon contents tend to be higher than the global average and increase as latitude increases within the temperate zones. Reductions in temperature and increases in waterlogging of soils at higher temperate latitudes are the main causes of this trend. These changes tend to enable a constant stream of biomass to enter the soils but reduce organic matter degradation. The global average estimated carbon density per unit area of soil in various vegetated zones and ecosystems are listed in Table 1.1.

1.3.1 Croplands

Soils underlying arable croplands typically possess low carbon soil contents (less than about 150 t C ha^{-1}), because they are either under cultivation or

Table 1.1 Estimated soil carbon complements in the top 1 m of common vegetated systems. The ranges provided indicate uncertainty in the global average carbon density for each major vegetated system considered (after Hester *et al.*, 2010) [14].

Ecosystem	Carbon density ($t\ C\ ha^{-1}$)
Wetland	643
Cropland (arable)	80–122
Tundra	127–206
Deserts and semi-deserts	42–57
Temperate grassland and shrubland	99–236
Tropical savanna and grassland	90–117
Boreal forest	247–344
Temperate forest	96–147
Tropical forest	122–123

have been cultivated in the past [14]. Typically, such soils are maintained under the condition that causes them to lose their carbon to the atmosphere and by leaching at a rapid rate. they present a major challenge for sustainable future sequestration of carbon in soils.

1.3.2 Grasslands

Grasslands typically display higher soil carbon contents rather than cropland soils. Table 1.1 displays the carbon density of undisturbed naturally occurring grassland soils. Some crop rotation schemes combine periods of grassland cultivation with periods of arable crop growth. Soils associated with such systems tend to have carbon densities higher than croplands but not as high as natural undisturbed grassland soils. The soil carbon density tends to increase for areas that are laid to grass for higher fractions of the crop rotation cycle. Extensive cultivation of fast-growing bio-energy grasses with deep root systems, such as switchgrass and miscanthus, has the potential to substantially increase the carbon density of grassland and cropland converted to grassland.

1.3.3 Woodlands

Woodlands consisting of conifer, deciduous or mixed tree types, particularly in high-latitude forests tend to have soils with carbon densities similar to, or somewhat higher than, undisturbed natural grasslands. Higher latitude coniferous forests tend to retain much higher soil carbon densities than deciduous temperate woodlands [41].

Sun *et al.* (2019) [22] evaluated the long-standing impacts of temperature, litter inputs, soil characteristics, and vegetation type on the carbon content in the top 20 cm of soils in forests along a climatic gradient in China content of the 0- to 20-cm soil layer and its fractions across three wet forested sites of contrasting climatic zones in China. They found that as mean annual temperature increased organic carbon in the soil samples reduced, whereas annual litterfall and soil microbial respiration increased. Carbon compounds displaying greater stability were better preserved in those forest soil samples from warmer climates. This was considered to be the most likely caused by the more rapid breakdown of less stable carbon molecules in warmer climate forest soils.

1.3.4 Temperate Wetlands and Peat Bogs

Temperate wetlands often associated with bogs, swamps, and extensive peat accumulations display the highest carbon densities of terrestrial soils. Peats can achieve carbon densities greater than 1000 t ha^{-1} [42], or 5 to 10 times the levels of the other vegetated zones described.

1.3.5 Induced Changes in Vegetation and Land Conditions

Most attempts to change the vegetation type on a land area, particularly from forests or natural undisturbed grasslands to croplands [43], lead to a reduction in soil carbon density. Land-use alteration from one type to another invariably involves soil disturbance that leads to more CO_2 lost from the soil to the atmosphere. Eventually, the modifications induced to vegetation type at the land surface will establish a new carbon equilibrium within the underlying soil. Of course, converting cropland to permanent grassland or woodland does ultimately lead to an increase in the carbon content of the underlying soils. In Europe, it is estimated that poorly managed and damaged cropland and gazing land with improved land management practices applied could offer additional carbon sequestration potential of between 0.1 and 0.6 t C ha^{-1} [14]. Based on the areal extent of such poorly managed land, the incremental soil carbon sequestration potential in Europe is estimated at between about 60 and 175 Tg C a^{-1}.

1.3.6 Warm Temperate and Tropical Vegetated Zones

The average carbon densities in forest soils of high-latitude Europe are known to be about three to five times higher than average carbon densities in soils of low-latitude Europe, with a 10-fold difference in their potential as carbon sinks [44]. On the other hand, with the exception of tropical peatlands, the carbon density of tropical soils is typically less than temperate soils and many times less than the vegetation that grows upon it. Soil carbon losses are high due to higher soil temperatures and water leaching rates. Potential carbon sequestration rates in tropical soils are about half those of temperate soils. Deforestation, a common practice in many tropical regions for conversion to cropland and grazing land, further exacerbates the relatively low carbon stock in tropical stores by releasing much of it rapidly to the atmosphere, especially where over grazing leads to drastic soil erosion.

1.4 Estimates of Global Potential for Carbon Sequestration in Soils

There are several published estimates of the global potential for carbon sequestration in soils, summarized by Paustian *et al.* (2019) [45], that suggest a range for the technical potential for additional carbon uptake by soils of between 2 and 5 Pg (gigatonnes) of CO_2 per year (~0.5 to 1.5 Pg C a^{-1}). To achieve the top end of that range, the best of currently available land and carbon management practices would need to be adopted for most croplands and grass lands around the world. These practices would also need to involve a drastic reduction in soil disturbance, measures to combat soil erosion, and be accompanied by a substantial increase in reforestation.

If and when such practices are adopted, they should provide an initial boost to the soils carbon uptake from the atmosphere. However, such a boost would likely only last for some decades until the better managed cropland and grassland soils and reforested regions achieved their new and higher carbon storage equilibrium. Technology advances in producing beneficial soil supplements cheaply, for example, biochar residues from biofuel production [46], and genetically engineered perennial and annual grain crops with deeper and larger roots [47], if widely utilized in the medium term, have the potential to increase the carbon uptake by soils by a further 0.5 to 1.0 Pg C a^{-1}. This could achieve up to about 2 Pg C a^{-1} of additional global soil carbon storage capacity [45].

1.5 Conclusions

The greatest potential to improve carbon sequestration rates in soils is by converting croplands into (or back into) woodlands, grasslands, and wetlands. Whereas intensive addition of farmyard manure, slurries, and general biomass waste into cropland soils can significantly increase their carbon densities, most intensive soil carbon intervention projects on croplands are less effective in the long term than converting such lands to natural forests, grasslands, and wetlands. Grassland and converted cropland carbon storage potential could be improved by more extensive growth of perennial deep-rooted, fast growing bioenergy crops such as switchgrass and miscanthus. Whereas, the top 2 m of soils currently hold about 2,400 Pg of carbon globally, there is a general consensus that improved land management practices could add to this by between 0.5 to 2.0 Pg C a^{-1}. A key limitation on the soils ability to rapidly uptake additional carbon from the atmosphere in coming decades is the long residence time of carbon in existing soils. This suggests that it would take significant time for the additional carbon added to soil from biomass to be converted and stabilized by carbon mineralization. Once carbon is mineralized it is more likely to be retained and sequestered long term in the ground rather than be cycled back into the atmosphere. Even in mineralized form, some soil carbon is frequently disturbed and displaced by the relentless forces of erosion, particularly over geological time scales. These factors mean that attempts to sequester more carbon in soils, while beneficial in the long term, can only constitute a relatively small component of a short-term global effort to sequester more carbon from the atmosphere to rapidly combat rising CO_2 levels in the atmosphere that are fueling climate change. However, efforts to prevent soil erosion and reduce soil disruption by agriculture are essential to preserve the carbon store currently held globally in soils to prevent as much as possible of that carbon being displaced to the atmosphere and contributing incrementally to CO_2 levels in the atmosphere.

References

1. Khajehesamedini, A., Sadatshojaie, A., Parvasi, P., Rahimpour, M.R., Naserimojarad, M.M., Experimental and theoretical study of crude oil pretreatment using low-frequency ultrasonic waves. *Ultrason. Sonochem.*, 48, 383–395, 2018.
2. Sadatshojaie, A. and Rahimpour, M.R., CO_2 emission and air pollution (volatile organic compounds, etc.)–related problems causing climate change, in: *Current*

Trends and Future Developments on (Bio-) Membranes, 1–30, Elsevier, https://www.sciencedirect.com/science/article/pii/B9780128167786000011, 2020.

3. Ghashghaei, S., Ghashghaei, S., Dastsooz, H., Fardaei, M., Haplotype analysis for five unrelated patients with a novel ATP7B gene mutation. *Eur. J. Hum. Genet., 26*, 1020–1020, https://scholar.google.com/scholar?hl=en&as_sdt=0,5&cluster=373618226633714896, 2018.
4. Ghashghaei, S., Sadatshojaei, E., Fardaei, M., Dastsooz, H., Wood, D.A., Identification of a Common Ancestor for 5 Unrelated Wilson Disease Patients with the Same ATP7B Mutation using SNP Haplotypes. *Eur. J. Hum. Genet.*, 2021.
5. Ghashghaei, S., Wood, D.A., Sadatshojaei, E., Jalilpoor, M., Accurate diagnosis of Covid-19 using artificial intelligence. *Sci. Rep.*, 2021.
6. Madani, F., Ghashghaei, S., Khashei, R., Sadatshojaei, E., Wood, D.A., Analysis of bacterial factors in relation to urinary tract infection and patterns of antibiotic susceptibility in recipients of kidney and liver transplants. *Sci. Rep.*, 2021.
7. Zare Sheibani, A., Ghashghaei, S., Kokabi, R., Mobasheri, F., Comparison of the effect of diclofenac suppository and lidocaine-prilocaine cream (XYLA cream) on pain relief after episiotomy in natural Childbirth. *Sci. Rep.*, 2021.
8. Zare, A., Ghashghaei, S., Poormantaseri, Z., Sharafi, M., Prevalence and factors related to arbitrary drug use in Fasa: Case study. *Br. J. Clin. Pharmacol.*, 2021.
9. Rezaei Dehshibi, R., Sadatshojaie, A., Mohebbi, A., Riazi, M., A new insight into pore body filling mechanism during waterflooding in a glass micromodel. *Chem. Eng. Res. Des.: Transactions of the Institution of Chemical Engineers Part A*, 151, 100–107, 2019.
10. Wilcox, J., Renforth, P., Kraxner, F. (eds.), *The Role of Negative Emission Technologies in Addressing Our Climate Goals*. Lausanne, Frontiers Media SA, https://scholar.google.com/scholar_url?url=https://www.frontiersin.org/research-topics/9752/pdf%3Futm_source%3Dfweb%26utm_medium%3Dnblog%26utm_campaign%3Dba-sci-ebook--20200300&hl=en&sa=T&oi=gsb-gga&ct=res&cd=0&d=16349358176583991556&ei=ROT-X9fXEYfPmAG-w14KwCw&scisig=AAGBfm1dUf9bJFQ1f8RVUxAXc4tQNKEVNQ, 2020.
11. Batjes, N.H., Total carbon and nitrogen in the soils of the world. *Eur. J. Soil Sci.*, 65, 1, 10–21, 2014.
12. Lal, R., Soil carbon stocks under present and future climate with specific reference to European ecoregions. *Nutr. Cycling Agroecosyst.*, 81, 2, 113–127, 2008.
13. Lal, R., Follett, R., Kimble, J., Achieving soil carbon sequestration in the United States: a challenge to the policy makers. *Soil Sci.*, 168, 12, 827–845, 2003.
14. Hester, R.E. and Harrison, R.M. (Eds.), *Carbon capture: Sequestration and storage* (Vol. 29). Royal Society of Chemistry, https://pubs.rsc.org/en/content/ebook/978-1-84755-917-3, 2010.
15. Anderson, M.C., PRO. *J. Gynecol. Surg.*, 7, 3, 191–194, 1991.

16. Schlesinger, W.H. and Andrews, J.A., Soil respiration and the global carbon cycle. *Biogeochemistry*, 48, 1, 7–20, 2000.
17. Hayes, M.H. and Clapp, C.E., Humic substances: considerations of compositions, aspects of structure, and environmental influences. *Soil Sci.*, 166, 11, 723–737, 2001.
18. Post, W.M. and Kwon, K.C., Soil carbon sequestration and land-use change: processes and potential. *Global Change Biol.*, 6, 3, 317–327, 2000.
19. He, Y., Trumbore, S.E., Torn, M.S., Harden, J.W., Vaughn, L.J., Allison, S.D., Randerson, J.T., Radiocarbon constraints imply reduced carbon uptake by soils during the 21st century. *Science*, 353, 6306, 1419–1424, 2016.
20. van Diemen, R., AI Annex I: Glossary, in: *Climate Change and Land*, p. 803, 2019.
21. Halldorsson, G., Sigurdsson, B.D., Finér, L., Gudmundsson, J., Kätterer, T., Singh, B.R., Vesterdal, L., Arnalds, A., *Soil Carbon Sequestration–for climate, food security and ecosystem services*, Nordic Council of Ministers, 2015.
22. Sun, X., Tang, Z., Ryan, M.G., You, Y., Sun, O.J., Changes in soil organic carbon contents and fractionations of forests along a climatic gradient in China. *For. Ecosyst.*, 6, 1, 1, 2019.
23. Alexandrov, G.A. and Matsunaga, T., Normative productivity of the global vegetation. *Carbon Balance Manage.*, 3, 1, 8, 2008.
24. Smith, J., Similar response of labile and resistant soil organic matter pools to changes in temperature. *Nature*, 433, 57–59, 2005.
25. Zhang, H. and Zhou, Z., Recalcitrant carbon controls the magnitude of soil organic matter mineralization in temperate forests of northern China. *For. Ecosyst.*, 5, 1, 17, 2018.
26. Paustian, K., Andren, O., Janzen, H., Lal, R., Smith, P., Tian, G., Tiessen, H., Van Noordwijk, M., Woomer, P., Agricultural soils as a sink to mitigate CO2 emissions. *Soil Use Manage.*, 13, 230–244, 1997.
27. Lal, R. and Kimble, J., Conservation tillage for carbon sequestration. *Nutr. Cycling Agroecosyst.*, 49, 1–3, 243–253, 1997.
28. Bellarby, J., Foereid, B., Hastings, A., *Cool Farming: Climate impacts of agriculture and mitigation potential*, Greenpeace, https://eprints.lancs.ac.uk/id/eprint/68831/1/1111.pdf, 2008.
29. Alvarez, R., A review of nitrogen fertilizer and conservation tillage effects on soil organic carbon storage. *Soil Use Manage.*, 21, 1, 38–52, 2005.
30. Lowe, A., Beasley, B., Berly, T., Carbon Capture and Storage (CCS) in Australia, in: *Issues in Environmental Science and Technology*, vol. 29, p. 65, 2010.
31. Ewing, S., Consuming online in Australia, in: *Urban consumption*, p. 105–122, 2011.
32. Feron, P.H.M., Attalla, M.I., Puxty, G., Allport, A., Cottrell, A.J., McGregor, J.A., Post-combustion capture (PCC) R&D and pilot plant operation in Australia. In: CSIRO Energy Technology, Newcastle (eds.) *IEA GHG 11th Post Combustion CO2 Capture Network meeting; 20–21 May 2008*, IEA

Greenhouse Gas R&D Programme, Vienna, Austria, http://hdl.handle.net/102.100.100/121594?index=1, 2008.

33. Conant, R.T., Easter, M., Paustian, K., Swan, A., Williams, S., Impacts of periodic tillage on soil C stocks: A synthesis. *Soil Tillage Res.*, 95, 1–2, 1–10, 2007.
34. Buhre, B.J., Elliott, L.K., Sheng, C., Gupta, R.P., Wall, T.F., Oxy-fuel combustion technology for coal-fired power generation. *Prog. Energy Combust. Sci.*, 31, 4, 283–307, 2005.
35. Hutchinson, J., The record of peat wastage in the East Anglian fenlands at Holme Post, 1848-1978 AD. *J. Ecol.*, 51, 68, 229–249, 1980.
36. Heathwaite, A.L., Disappearing peat-regenerating peat? The impact of climate change on British peatlands. *Geog. J.*, 159, 203–208, 1993.
37. Lal, R., Climate change and food security soil carbon sequestration impacts. *Science*, 304, 1623, 2004.
38. Lal, R., *Soil erosion and carbon dynamics*, Elsevier, https://www.sciencedirect.com/science/article/pii/S0167198704001898, 2005.
39. Wardle, D.A., Islands as model systems for understanding how species affect ecosystem properties. *J. Biogeogr.*, 29, 5–6, 583–591, 2002.
40. Holden, J., Shotbolt, L., Bonn, A., Burt, T., Chapman, P., Dougill, A., Fraser, E., Hubacek, K., Irvine, B., Kirkby, M., Environmental change in moorland landscapes. *Earth-Sci. Rev.*, 82, 1–2, 75–100, 2007.
41. Harrison, A., Howard, P., Howard, D., Howard, D., Hornung, M., Carbon storage in forest soils. *Forestry: An International Journal of Forest Research*, 68, 4, 335–348, 1995.
42. Chapman, S., Bell, J., Donnelly, D., Lilly, A., Carbon stocks in Scottish peatlands. *Soil Use Manage.*, 25, 2, 105–112, 2009.
43. Burges, C.J., A tutorial on support vector machines for pattern recognition. *Data Min. Knowl. Discovery*, 2, 2, 121–167, 1998.
44. Liski, J., Perruchoud, D., Karjalainen, T., Increasing carbon stocks in the forest soils of western Europe. *For. Ecol. Manage.*, 169, 1–2, 159–175, 2002.
45. Paustian, K., Larson, E., Kent, J., Marx, E., Swan, A., Soil C sequestration as a biological negative emission strategy, in: *The Role of Negative Emission Technologies in Addressing Our Climate Goals*, 2020.
46. Song, X., Pan, G., Zhang, C., Zhang, L., Wang, H., Effects of biochar application on fluxes of three biogenic greenhouse gases: a meta-analysis. *Ecosyst. Health Sustainability*, 2, 2, e01202, 2016.
47. Crews, T.E. and Rumsey, B.E., What agriculture can learn from native ecosystems in building soil organic matter: A review. *Sustainability*, 9, 4, 578, 2017.

2

A Brief Insight on Factors Controlling Rate of Chemical Weathering of Minerals Existing in Soil

Tejaswini Sahoo[1], Rashmirekha Tripathy[1], Jagannath Panda[1,3], Madhuri Hembram[1], Saraswati Soren[1], Deepak Kumar Senapati[1], C.K. Rath[1], Sunil Kumar Sahoo[2] and Rojalin Sahu[1*]

[1]*School of Applied Sciences, Kalinga Institute of Industrial Technology, Deemed to be University, Bhubaneswar, India*
[2]*Health Physics Division, BARC, Mumbai, India*
[3]*CSIR-IMMT, Bhubaneswar, India*

Abstract
Mineral weathering relative to soil has two parts: weathering happened previously where hard rocks have been broken down into clay which ultimately forms soil, loams, and unconsolidated sands both chemically and physically and other is soil and mineral weathering happening currently which acts as vital source of crop nutrients. This chapter discusses background of chemical weathering of minerals, sequence of weathering of minerals from soil, mainly throwing light on the factors which controls the rate of chemical weathering like temperature and time factor, biotic process, oxidation, reduction, water, leaching, acidity, and many more.

Keywords: Chemical weathering, time factor, temperature factor, leaching, oxidation, reduction

2.1 Introduction

Weathering of rocks can be defined as the modification in the rate of composition and consolidation, occurring in the crust of earth under the influence of hydrospheric and atmospheric factors [1]. It has two different

**Corresponding author*: rsahufch@kiit.ac.in

Inamuddin, Mohd Imran Ahamed, Rajender Boddula and Tariq Altalhi (eds.) Applied Soil Chemistry, (23–40) © 2021 Scrivener Publishing LLC

categories: chemical and physical. Physical weathering is defined as formation of unconsolidated state of rock from consolidated rock. In case of chemical weathering, there is chemical composition modification either in the unconsolidated or consolidated rock. Merrill *et al.*, in the year 1906 [2–4], elaborated weathering as insincere changes caused in masses of rock due to atmospheric conditions, thereby causing complete or partial damage of the rock.

Physical weathering was referred as disintegration, whereas chemical weathering referred to as decomposition. As already discussed, in case of physical weathering, large masses of rock get converted into elastic or unconsolidated state, owing to the hydrospheric and atmospheric factors. Polynov *et al.*, in the year 1937 [5–7], states that weathering is a cyclic, complicated procedure which involves breakdown of solid structure of rock and production of enhanced interfacial interaction between the rock and nearby surrounding.

By mechanical conditions, in case of disintegration or physical weathering process, rocks break down into smaller form which retain both its composition as well its originality. For example, even though basalt gets broken down via physical weathering or mechanical means into its smaller form, it could be identified as basalt. Physical weathering of rock occurs either in place or by the media like wind, gravity displacement at steep slopes, moving water and glacial ice which carries the rock and then grind it. In place physical weathering is driven by five factors as stated by Rieche *et al.* in the year 1950 [8–10]. The factors include unloading, thermal expansion, crystal growth, colloid plucking, and pressure and abrasion due to animals and plants. When there is reduction in confining pressure due to joints, cracks, erosion, and uplift, unloading causes rock masses expansion. According to Polynov *et al.*, in the year 1937 [5], thermal contraction and expansion is vital factor, but as per Origgs *et al.* and Blaclrwelder *et al.* [11, 12], it is comparatively not important. Crystal growth causes production of prying action on minerals and rocks. This involves activity due to frost and also up to some extent formation of crystals due to chemical weathering, for example, as per Humbert and Marshall, in the year 1943 [13–15], quartz physical weathering occurred due to films of iron oxides. Colloid plucking refers to the physical damage occurred owing to the reduction of colloidal matter, resulting in peeling of the rock surface. According to Kellog *et al.*, in the year 1943 [16–18], destruction of rocks occurs due to overgrowth of roots of the plants.

On the other hand, chemical weathering refers to modifications in the rocks' chemical properties, owing to the hydrospheric and atmospheric factors. It is also known as the process of decomposition of minerals and rocks. The substance formed after the process of chemical weathering must include new mineral (mechanism of synthesis) or after the weathering the products should exist as residue after the elimination of other constituents (mechanism of residue). According to Polynov *et al.*, in the year 1937, the product obtained as per mechanism of synthesis is known as products of accumulation and the product obtained from mechanism of residue is known as ortho eluvium. Moreover, the solutes getting washed and accumulating in ground water are also considered as weathering products, but in the context of mineral content of soil, it is of least importance. This chapter discusses background of chemical weathering of minerals, sequence of weathering of minerals from soil, mainly throwing light on the factors which controls the rate of chemical weathering like temperature and time factor, biotic process, oxidation, reduction, water, leaching, acidity, and many more.

2.1.1 Weathering Similar to Hydrothermal and Diuretic Alteration of Minerals

Some minerals and layer silicates also form part of weathering research which are obtained from hydrothermal and deuteric changes. Layer silicates, consequently, appear at more depth of geologic column in comparison to the weathering depths during initiation. Even though silicates find its way to soil during the exposure of formation of rock due to erosion at the surface, they does not represent example of c weathering process chemically. According to Bowen *et al.*, in the year 1922 [19–21], occurrence of deuteric changes was observed later during rock crystallization. According to Harker *et al.*, in the year 1932 [22–24], during the rock metamorphosis, from feldspar, sericite was formed and chlorite might formed from garnet biotite schist. Shand *et al.*, in the year 1944 [25–27], reported that hydrothermal alteration is modifications in rock occurring due to the hot solution movement through specified routes into the earth. Some examples of this type of weathering were reported by Meyer and Sales *et al.*, in the year 1950 [1], sericite formation from hydrothermal alteration of quartz monazite up to 20 ft from the channel along which hot hydrothermal solution travelled. Kaolinite was formed further up to another 20 ft. Furthermore, hydrothermal alteration up to 60 ft formed montomorillonite. Variations in the rate of alterations depended on the distance of fracture or channels from the hydrothermal solutions. The examples explained above come in the category of alterations of rock owing to the movement of solutions in the depth of the earth. Moreover, according to Schmitt *et al.*, in the year 1950 [1], there are cases of same kind of alterations arising due to movement of hydrothermal solutions close to the surface. Water of hot spring at Yellowstone Park altered rocks (volcanic) into calcite and zeolites, orthoclase and quartz, and kaolinite and beidellite containing rocks.

2.2 Comparitive Stability of Minerals on the Basis of Their Sequence of Weathering

Depending on the factors initiating the chemical weathering process, weathering of minerals depends on their stability. Owing to the stability some gets weathered slower and some faster in comparison to other minerals. The process of weathering of minerals depending upon their stability is occurring in a sequence, often referred as weathering sequence. Establishment of sequence of weathering process occurs on the basis of certain criteria, comparative persistence with depth of formation, age of formation, size of particle, intensity factors of weathering, and geographical factors.

2.2.1 Heavy Minerals

According to Pettijohn *et al.*, in the year 1941 [28–30], reported rocks whose age was increasing included decreasing quantity of higher specific gravity minerals having less stability. Hence, among rocks which are older in comparison to Pleistocene, olivine was not included. The weathering sequence given by Pettijohn in the order of increasing order of stability includes (most unstable mineral first), olivine (22), actinolite (21), diopside (20), hypersthene (19), sillimanite (18), augite (17), zoisite (16), sohene (15), topaz (14), andalusite (13), hornblende (12), epidote (11), kyanite (10), staurolite (9), magnetite (8), ilmenite (7), apatite (6), biotite (5), garnet (4), monazite (3), tourmaline (2), zircon (1), rutile (–1), muscovite (–2), and anatade (–3). The last three minerals with negative numbers shows formation tendency instead of disappearance at the time of prolonged burial. According to Weyl *et al.*, in the year 1952 [1], it was suggested to classify four groups of heavy minerals according to their relative stability. The very stable group included magnetite titanite, rutile, zircon, and tourmaline. The slightly stable group included epidote and garnet. The most unstable minerals included augite, hot blende, and olivine. The above discussion shows way how heavy minerals stability could be used for establishing rock formation age, out of which soil was formed.

2.2.2 Coarsely Grinded Minerals

According to Goldich *et al.*, in the year 1938 [31–33], stability sequence of coarsely grained minerals was reported as (most stable first) quartz, muscovite, orthoclase, plagioclase, biotite, amphibole, and pyroxene. Marel *et al.*, in the years 1949, 1948a, and 1947 [34–36] reported magmatic minerals stability, keeping the particle size and weathering conditions similar (most resistant first), quartz, tourmaline, rutile, staurolite, zircon, allanite, albite, microcline, orthoclase, garnet, muscovite, oligodase, andesine, bytownite, epidote, anorthite, ampjibole, augite, biotite, hypersthene, olivine, and volcanic glass. As coarsely grinded quartz is not susceptible toward weathering, it is considered as stable among all the minerals.

2.2.3 Clay Size Mineral Particles

When the size of the particles of mineral becomes finer, then the sequence of stability of mineral becomes different relative to the coarsely grinded particle of mineral. This happens due to the enhancement of specific surface which increases the process of mineral weathering, which was comparatively more stable in coarsely grinded particle size. The particle size of clay minerals is below 200 diameters. According to Jackson *et al.*, in the year 1948 and 1952 [37–39], the sequence of weathering consisted of 13 stages for clay size particles. The sequence is in descending order of stability:

- Anatase (corundum, leucoxene, ilmenite, rutile, zircon, etc.)
- Hematite (limonite, goethite, etc.)
- Gibbsite (also allophane, boehmite, etc.)
- Kaolinite
- Montmorillonite (also saponite, beidellite, etc.)
- Vermiculite and silicates interstratified 1:2 layer
- Muscovite (also illite, sericite, etc.)
- Quartz
- Albite (also orthoclase, microline, stilbite, anorthite, etc.)
- Biotite (also nontronite, antigorite, magnesium chlorite, glauconite, etc.)
- Olivine-hornblende (also diopside, pyroxenes, etc.)
- Calcite (also apatite, aragonite, dolomite, etc.)
- Gypsum (also ammonium chloride, sodium nitrate, halite, etc.)

There are few modifications in the above list. Apatite was included at calcite stage; allophane was included in gibbsite stage by Tamura *et al.*, in the year 1953 [40–42]. Marel *et al.*, in the year 1947, and Haseman and Marshall, in the year 1943 [43–45], included zircon in Anatase stage. Woof and Carroll *et al.*, in the year 1951 [46–48], and Marsden and Tyler, in the year 1938 [49–51], included leucoxene in the Anatase stage.

2.3 Factors Affecting the Rate of Chemical Weathering

The rate of reaction involved in a chemical weathering is under the control wide range of capacity and intensity factors, acting as function of integration of capacity, intensity, and time. The weathering time can give rise to stage or degree of weathering. The various driving force affecting the rate of reaction involved in chemical weathering can be recognized to a extent of consideration for formation of soil are five usual factors including, time, parent material, relief, biotic forces, and climate. The conditions affecting the reaction rate of chemical weathering needs to be analyze more specifically in categories other than above listed five factors for formation of soil. Leaching effect is included in single intensity factor whether it is under the control of internal drainage, relief, rate of evaporation, distribution of rainfall, and amount of rainfall. The extent as well as the nature of leaching is collectively significant for determining the process of chemical weathering. Reduction and oxidation are specifically considered whether controlled by valence of the ions in the minerals, texture of the material, etc.

2.3.1 Capacity Factors Which Controls the Reaction Rate of Chemical Weathering

2.3.1.1 Specific Surface Role

The higher the specific surface area of the targeted mineral, the finer the size of the particle and hence very speedily it gets affected by chemical weathering processes as reported by Polynov *et al.*, in the year 1937 [5], and Merrill *et al.*, in the year 1906 [2]. The enhancement of the specific surface is inversely proportional to the material particle diameter. It has been reported that sequence of weathering are different in case of particle with fine size in comparison to that of coarsely sized particles. For example, due to effect of specific surface the rate of weathering of volcanic ash is faster in comparison to lava. This happens due to the larger surface area subjected for processes for weathering.

2.3.1.2 *Specific Weatherability Role of Mineral*

According Jackson *et al.*, in the year 1948 [37], it was reported that the chemical weathering rate is dependent on the mineral specific nature. Minerals specific nature is highly significant for determination of rate and course of weathering, which can be arranged as per the stability of the minerals. Searle *et al.*, in the year 1923 [52–53], reported that nature of rock significantly affected the weathering rate. As per Stephens *et al.* [54–56], depending on the weathering ease, there is no such considerable rock classification, with an exception for basalt which is considered as one of the rock which undergoes weathering very easily and the presence of high amount of silica causes slow weathering process. According to specific minerals, feldspars undergoes chemical weathering faster than quartz and slower than ferromagnetic minerals; calcite is more resistant to chemical weathering in comparison to gypsum. According to Goldich *et al.*, in the year 1938 [38], it was reported that plagioclase undergoes chemical weathering faster than potassium feldspar.

2.3.2 Intensity Factors Which Drives the Chemical Weathering Reaction Rate

2.3.2.1 *Factor of Temperature*

The contribution of temperature in controlling the rate of reaction of weathering is identified by difference in the rate of reaction in the temperate and tropical regions. Other factors of intensity like accumulation of organic matter, hydrolysis, and leaching operation are also affected by the temperature. Stephens *et al.*, in the year 1949 [54–56], reported that the air temperature index given by Szymkiewicz *et al.* in the year 1947 [57–59], T = 2.5t/10, where t represents the temperature of air in degree centigrade that is related to the chemical weathering rate. The significance of temperature on the rate of reaction involved in chemical weathering is shown by accomplishment of latosolization even on rocks which are acidic in nature, subject to condition intensity factor are sufficiently large. According to Williams *et al.*, in the year 1949. [60–62], it was reported that minerals chemical weathering is highly comprehensive in cool temperature regions, also in presence of snow cover. Retzer *et al.*, in the year 1949

[63–65], reported mineral colloids formation in the Rocky Mountains old soils, where cold temperature condition exists.

2.3.2.2 Leaching and Water Contribution

Water has two unique roles as a chemical weathering agent. Firstly, water acts as a reactant in hydroxylation, i.e., hydrolysis reaction which is vital source for accumulated hydroxyl in feralitic hydroxyl oxides and layer silicates. The reaction mentioned above is exothermic in nature. Minerals hydration gives rise to enhancement in volume and hence acts as a significant factor in coarsely grained igneous rocks decaying, as reported by Rieche *et al.*, in the year 1950 [8]. The second contribution of water is its role as leaching agent in internal drainage and flow of water allows the movement through the soil. The processes, leaching and hydroxylation, are connected to the extent that the hydrolysis products which are soluble should be eliminated for driving the reaction. Kellogg *et al.*, in the year 1943 [16], reported that water is main agent of chemical weathering. Bayers *et al.*, in the year 1933 [66–68], reported that the process of weathering completely expressed as hydrolysis reaction, but they also marked that the rate of process is affected by factors like specific nature of minerals, size of particles, movement of water, temperature, etc. Marbut *et al.*, in the year 1951 [69–71], reported that rock decomposition product contains variably large quantity of combined water in comparison to materials which are not decomposed. Mineralogical analysis is the basis of hydroxyl occurrence, by direct hydroxyl determination as per Jackson and Evans *et al.*, in the year 1952 [72–74], by weight loss as per Nutting *et al.*, in the year 1943 [75–77], by differential thermal methods as per Rowland and Grim *et al.*, in the year 1942 [78–80]. As per Jackson *et al.*, in the year 1948 [37], it was reported that chemical weathering is driven by leaching which involves the elimination of few products which are formed and moreover ions leaching might be enough comprehensive that upcoming stages of weathering consists of varying elements chemically. Another important factor of leaching is calculation of solutes present in water of river. Polynov *et al.*, in the year 1937 [5], and Smith *et al.*, in the year 1913 [93–95], reported mobility of ions in comparative manner to describe abovementioned calculation. The order of ions in the increasing order includes HO_3, SiO_2, Mg, Na, and Ca. Further, by extending this study, Goldich *et al.*, in the year 1938 [31], reported according to increasing order of loss, SiO_2, I, Mg, Ca, and Na. It was considered that Al was neither lost or accumulated and combined water and iron was gained. Mohr *et al.*, in the year 1944 [96–98], reported the significance of internal drainage due to leaching and reactions of weathering. The motion of water in the soil is controlled by drainage. As per Jackson *et al.*, in the year 1948 [38], sediments of tourmaline, zircon, quartz, feldspar, and authigenic micas indicated reversal in a case when the ionic substances were abundantly existed in sea water.

2.3.2.3 *Acidity Factor*

Material pH value and water moving through it has a considerable role in the nature and rate of chemical weathering. As per Jackson *et al.*, in the year 1948 [38], it was reported that the enhancement in acidity has a positive contribution in the rate function of weathering. Graham *et al.*, in the year 1941b [81–83], reported feldspar faster decomposition when acidified clay is present. The procedure shows faster elimination of cations (metallic) by exchange materials and simultaneously hydrogen is released during the process of hydroxylation reaction as the product of weathering. Additionally, with acid saturation exchange, also the influence of specific acid likes sulphuric and carbonic acid should be under consideration. Truog and Attoe *et al.*, in the year 1946 [84–86], reported release of potassium dilute acid from soil. Loss of mica and depotassication, with increment in kaolinite and vermiculite are highly enhanced with the arrival of leaching, which lowers the pH of the soil lower to 7, as reported by Jackson and Hseung in the year 1952 [87–89]. Gibbsite and kaolinite are formed because of the enhancement of the acidic condition of soil, in case of minerals containing high quantity of silica. This was explained by Hendricks and Ross *et al.* [90–92], in the year 1945, as high rate of elimination of silica and metallic ions due to leaching prior to recombination with sesquioxides. Even though, boehmite, gibbsite, and kaolinite are structurally hydroxyl compounds, but they act chemically like acid anhydrides or insoluble weak acids, weaker in comparison to silicic acid.

2.3.2.4 *Biotic Processes Factor*

Biotic processes in case of chemical weathering can be discussed in two sections. One is organic residue effects and other is ionic uptake cycles. Plants which are very tall physically affects chemical weathering by increasing the size of the cracks in rocks, as a result of which movement of water is easier and also it shades the soil, thereby affecting evaporation and soil temperature. High plants as well as microorganisms fasten the nutrients release from minerals through exchange of hydrogen ions with minerals charge and absorption. According to Bastisse and Demolon *et al.* in the year 1946 [1], it was reported that plants were one of responsible factors of weathering owing to the fact that soil planted with trees released higher quantity of potassium in comparison to unplanted soil. Jackson and Hseung *et al.*, in the year 1952 [87], reported that soil acidity accumulation is followed by micas depotassication. As per Kellog *et al.*, in the year 1943 [16], it was reported that in case of bare soil or virgin soil due to vegetation nutrients are returned as organic residue to the soil surface. Organic matter accumulation on the surface of the soil has considerable impact on the minerals weathering, either by distribution of aluminium oxide and iron present in soil or by bases leaching. As per Kanehiro and Sherman *et al.* in the year 1948 [99–101], it was reported that ferns present in regions of tropical humid Hawaii, below highly acidic forest floor have reactions in the pH range between 3 and 4,which is relative to the white pine forest floor the zone of cool temperate region. The organic matter mixed with soil controls the rate of chemical weathering and loss of bases. The rate of chemical weathering is faster in case of highly acidic forest floor in comparison to alkaline or neutral forest floor. Some kinds of organic matter, specifically organic acids, give rise to formation of complex with sesquioxide ions and transfer them to subsoil from the upper horizon and little amount inside the ground water. Elimination of iron out of ferromagnesian minerals is increased owing to this activity. The overall process of weathering as reported by Jackson and Hseung in the year 1952 [16] was influenced because of elimination of iron oxides from horizon A and accumulation in horizon B. As per Vagelor *et al.*, in the year 1933 [102–104], and Mohr *et al.*, in the year 1944 [96], it was reported that in tropical areas vegetal canopy is highly significant. According to Mohr *et al.*, in the year 1944, it was reported that air temperature above the soil is considerably less in case of canopy forest. Similarly, soil enveloped by forest has 10°C to 15°C. The decrease in the temperature makes all the chemical reaction slower. Organic matter decomposition is slower, causing deposition of organic substances either in or at horizon surface. The condition in deciduous tropical forest and evergreen tropical forest is considerably different. Gupta and Griffith *et al.*, in the year 1947 [105], reported that when teak is planted in laterite soil then a crust of laterite is formed due to the soil dehydration during the dormant leafless stage or drought period. The laterite soil formed hampers the tree growth. Sherman *et al.*, in the year 1953 [106], verified these findings in case of Hawaiian Islands.

2.3.2.5 *Reduction and Oxidation Factor*

Reduction and oxidation reaction influences the process of chemical weathering. As per Jackson *et al.* in the year 1948 [37], in equation of weathering, positive sign is assigned to oxidation. Hence, iron occurrence in ferrous state or reduction favoring conditions enabled the formation of montmorin in the intermediate stage of weathering as per Wendricks and Ross in the year 1945 [107]. Moreover, occurrence of iron in ferric state or oxidation favoring conditions favored free oxides and kaolin formation in higher weathering stage. According to Marshall and Humbert *et al.*, in the year 1943 [13], oxidation role in weathering was reported as ferric oxide formation in quartz grains cracks which breaks the quartz crystal. Oxidation rate also has an impact on the accumulated organic matter prevailing in soil which ultimately influences the rate of chemical weathering.

2.3.2.6 Influence of Time on the Rate of Chemical Weathering

According Jackson *et al.*, in the year 1948 [37], it was reported that for a particular particle size range and mineral species, the stages of weathering of the colloid weathering are the result of time duration of weathering and intensity function multiplication. Thorp *et al.*, in the year 1944 [108], reported that soils present in some of the mountain displayed considerable development of soil and weathering after a particular duration of time. Retzer *et al.*, in the year 1949 [63], reported great significance of the age of the mineral weathering and formation of dense, unique subsoils which are oftenly highly thick in case of Rocky Mountains of Central Colorado. The time of exposure or age exhibits crucial part for determining the type of mineral existing at a particular time in tropical soils. Mohr *et al.*, in the year 1944 [96], reported five stages of weathering for the development of tropical soils. The stages include the following:

- Beginning stage (unweathered virgin material)
- Juvenile stage (point where there is beginning of weathering)
- Virile stage (in this stage, the weathering is in highly advanced stage where the percentage of unweathered material is negligible)
- Senile stage (at this point, there is completion of unweathered material weathering)
- Final or end stage (at this point, development has reached to completion and the soil has been weathered.)

The above proposal displays that the weathering of mineral which contributes in formation of soil remains in equilibrium (dynamic) with the factors prevailing in the environment. According to the above proposal, at the final stage, there is the establishment of static equilibrium.

2.4 Conclusion

In this chapter, a brief summarization of process of chemical weathering and various factors like specific surface area, oxidation and reduction, biotic process, temperature, time, acidity, leaching, effect of water, etc., responsible for controlling the rate of chemical weathering is discussed.

References

1. Jackson, M.L. and Donald Sherman, G., Chemical weathering of minerals in soils, in: *Advances in agronomy*, vol. 5, pp. 219–318, Academic Press, Cambridge, Massachusetts, 1953.
2. Ruxton, B.P., Measures of the degree of chemical weathering of rocks. *J. Geol.*, 76, 5, 518–527, 1968.
3. Reiche, P., Graphic representation of chemical weathering. *J. Sediment. Res.*, 13, 2, 58–68, 1943.
4. Parker, A., An index of weathering for silicate rocks. *Geol. Mag.*, 107, 6, 501–504, 1970.
5. Polynov, B.B., The Cycle of Weathering, translated from the Russian by A, in: *Muir, with a foreword by WG Ogg*, Thomas Murby, London, 1937.
6. Mill, H.R., Geography: principles and progress, in: *The international geography*, pp. 2–13, 1899.
7. Polynov, B.B., Types of weathering crust, in: *Transactions of the Third International Congress of Soil Science*, Oxford, England, vol. 1, pp. 327–330, 1935.
8. Reiche, P., *A survey of weathering processes and products*, Univ. New Mexico Geol. Publ. No. 3, New Mexico, 1950.
9. Keller, W.D., The principles of chemical weathering. *LWW*, 80, 2, 165, 1955.
10. Goldich, S.S., A study in rock-weathering. *J. Geol.*, 46, 1, 17–58, 1938.
11. Blackwelder, E., The insolation hyphothesis of rock weathering. *Am. J. Sci.*, 152, 97–113, 1933.
12. Griggs, D.T., The factor of fatigue in rock exfoliation. *J. Geol.*, 44, 7, 783–796, 1936.
13. Humbert, R.P. and Marshall, C.E., *Mineralogical and chemical studies of soil formation from acid and basic igneous rocks in Missouri*, University of Missouri, College of Agriculture, Agricultural Experiment Station, Missouri, 1943.
14. Cady, J.G., Rock weathering and soil formation in the North Carolina Piedmont region, in: *Soil Sci. Soc. Am. Proc*, vol. 15, pp. 337–342, 1950.
15. Allen, V.T., Weathering and heavy minerals. *J. Sediment. Res.*, 18, 1, 38–42, 1948.

16. Kellogg, C.E., *The soils that support us,* vol. 52, no. 5, LWW, Philadelphia, 1941.
17. Shepard, W., Food or famine: the challenge of erosion, in: *Food or famine: the challenge of erosion.*
18. Gustafson, A.F., *Conservation of the Soil,* McGraw-Hill Book Company, Inc., New York, 1937.
19. Bowen, N.L., The reaction principle in petrogenesis. *J. Geol.*, 30, 3, 177–198, 1922.
20. Morey, G.W., Experimental geology. *J. Phys. Chem.*, 60, 6, 718–724, 1956.
21. Ito, K.E., II and Kennedy, J.C., Experimental study of basalt-garnet granulite–eclogite. *Geokhimiya,* 10, 4, 415–427, 1972.
22. Higgins, M.W., *The geology of the Brevard lineament near Atlanta,* Georgia. No. 77, Georgia State Division of Conservation, Department of Mines, Mining and Geology, Georgia, 1966.
23. Reed Jr., J.C., Bryant, B., Myers, W.B., The Brevard zone: A reinterpretation, in: *Studies in Appalachian Geology: Central and Southern*, pp. 261–269, Interscience, New York, 1970.
24. Crawford, T.J. and Medlin, J.H., The western Georgia Piedmont between the Cartersville and Brevard fault zones. *Am. J. Sci.*, 273, 8, 712–722, 1973.
25. Shand, S.J., The terminology of late-magmatic and post-magmatic processes. *J. Geol.*, 52, 5, 342–350, 1944.
26. Visher, S.S., Geologists Starred, 1903-43: Where Educated, Age. *J. Geol.*, 53, 5, 326–336, 1945.
27. Chhibber, H.L., Origin of the bauxite deposits of India. *J. Sci. Ind. Res., Delhi*, 5, 4, 48–51, 1946.
28. Pettijohn, F.J., Persistence of heavy minerals and geologic age. *J. Geol.*, 49, 6, 610–625, 1941.
29. Dryden, A.L. and Dryden, C., Comparative rates of weathering of some common heavy minerals. *J. Sediment. Res.*, 16, 3, 91–96, 1946.
30. Morton, A.C., Stability of detrital heavy minerals in Tertiary sandstones from the North Sea Basin. *Clay Miner.*, 19, 3, 287–308, 1984.
31. Goldich, S.S., A study in rock-weathering. *J. Geol.*, 46, 1, 17–58, 1938.
32. Wolf, A., Weathering of the Medford Diabase: Pre-or Postglacial? *J. Geol.*, 40, 5, 459–465, 1932.
33. Velbel, M.A., Geochemical mass balances and weathering rates in forested watersheds of the southern Blue Ridge. *Am. J. Sci.*, 285, 10, 904–930, 1985.
34. Van Der Marel, H.W., Tropical soils in relation to plant nutrition. *Soil Sci.*, 64, 6, 445–452, 1947.
35. Van der Marel, H.W., Gamma ferric oxide in sediments. *J. Sediment. Res.*, 21, 1, 12–21, 1951.
36. Van Der Marel, H.W., Quantitative differential thermal analy ses of clay and other minerals. *Am. Min.: Journal of Earth and Planetary Materials*, 41, 3–4, 222–244, 1956.

37. Jackson, M.L., Tyler, S.A., Willis, A.L., Bourbeau, G.A., Pennington, R.P., Weathering sequence of clay-size minerals in soils and sediments. I. Fundamental generalizations. *J. Phys. Chem.*, 52, 7, 1237–1260, 1948.
38. Jackson, M.L., Tyler, B., Willis, A.L., Bourbeau, G.A., Pennington, R.P., Weathering sequence of clay-size minerals in soils and sediments: I. *Fundamental generalization. J. Phys. Colloid Chem.*, 52, 1237–1260, 1948.
39. Jackson, M.L., Hseung, Y., Corey, R.B., Evans, E.J., Vanden Heuvel, R.C., Weathering Sequence of Clay-size Minerals in Soils and Sediments: II. Chemical Weathering of Layer Silicates 1. *Soil Sci. Soc. Am. J.*, 16, 1, 3–6, 1952.
40. Tamura, T., Jackson, M.L., Sherman, G.D., Mineral Content of Low Humic, Humic and Hydrol Humic Latosols of Hawaii 1. *Soil Sci. Soc. Am. J.*, 17, 4, 343–346, 1953.
41. Jackson, M.L. and Donald Sherman, G., Chemical weathering of minerals in soils, in: *Advances in agronomy*, vol. 5, pp. 219–318, Academic Press, Cambridge, Massachusetts, 1953.
42. Jackson, M.L., Wm Levelt, T.H., Syers, J.K., Rex, R.W., Clayton, R.N., Sherman, G.D., Uehara, G., Geomorphological Relationships of Tropospherically Derived Quartz in the Soils of the Hawaiian Islands 1. *Soil Sci. Soc. Am. J.*, 35, 4, 515–525, 1971.
43. Haseman, J.F. and Marshall, C.E., *The use of heavy minerals in studies of the origin and development of soils*, University of Missouri, College of Agriculture, Agricultural Experiment Station, Missouri, 1945.
44. Marshall, C.E. and Haseman, J.F., The quantitative evaluation of soil formation and development by heavy mineral studies: a Grundy silt loam profile. *Soil Sci. Soc. Am. J.*, 7, C, 448–453, 1943.
45. Haseman, J.F. and Marshall, C.E., The Use of Heavy Minerals in Studies of the Origin and Development of Soils. *Missouri Agricultural Experimental Station. Res. Bull.*, 387, 387, 1945.
46. Carroll, D. and Woof, M., Laterite developed on basalt at Inverell, New South Wales. *Soil Sci.*, 72, 2, 87–100, 1951.
47. Carroll, D., Description of a Montalto soil in Maryland. *Soil Sci.*, 75, 2, 87–102, 1953.
48. Carroll, D. and Jones, N.K., Laterite developed on acid rocks in southwestern Australia. *Soil Sci.*, 64, 1, 1–16, 1947.
49. Tyler, S.A. and Marsden, R.W., The nature of leucoxene. *J. Sediment. Res.*, 8, 2, 55–58, 1938.
50. Bailey, S.W. and Cameron, E.N., Is leucoxene always finely crystalline rutile?; reply. *Econ. Geol.*, 52, 6, 716–720, 1957.
51. Allen, V.T., Is leucoxene always finely crystalline rutile?; discussion. *Econ. Geol.*, 51, 8, 830, 1956.
52. Pettijohn, F.J., Persistence of heavy minerals and geologic age. *J. Geol.*, 49, 6, 610–625, 1941.
53. Jackson, M.L., Clay transformations in soil genesis during the Quaternary. *Soil Sci.*, 99, 1, 15–22, 1965.

54. Stephens, C.G., Comparative morphology and genetic relationships of certain Australian, North American and European soils. *J. Soil Sci.*, 1, 123–1495, 1950.
55. Hallsworth, E.G. and Costin, A.B., Soil classification. *J. Aust. Inst. Agric. Sci.*, 16, 84–895, 1950.
56. Nagelschmidt, G., Desai, A.D., Muir, A., The minerals in the clay fractions of a black cotton soil and a red earth from Hyderabad, Deccan State, India. *J. Agric. Sci.*, 30, 4, 639–653, 1940.
57. Prescott, J.A., A climatic index for the leaching factor in soil formation. *J. Soil Sci.*, 1, 1, 9–19, 1950.
58. Hosking, J.S., The soil clay mineralogy of some Australian soils developed on granitic and basaltic parent material. *J. Counc. Sci. Ind. Res., Australia*, 13, 206–2165, 1940.
59. Kelley, W.P., Dore, W.H., Woodford, A.O., Brown, S.M., The colloidal constituents of California soils. *Soil Sci.*, 48, 3, 201–256, 1939.
60. Williams, J.E., Chemical weathering at low temperatures. *Geogr. Rev.*, 39, 1, 129–135, 1949.
61. McCabe, L.H., Nivation and corrie erosion in West Spitsbergen. *Geog. J.*, 94, 6, 447–465, 1939.
62. Ekblaw, W.E., The importance of nivation as an erosive factor, and of soil flow as a transporting agency, in northern Greenland. *Proc. Natl. Acad. Sci. U.S.A.*, 4, 9, 288, 1918.
63. Fieldes, M., Chemical weathering of silicates in soil formation. *New Zealand J. Sci. Technol., B*, 36, 140–154, 1954.
64. Fieldes, M. and Williamson, K., II, Clay mineralogy of New Zealand soils; part I, Electron micrography. *New Zealand J. Sci. Technol.*, 37, 314–3355, 1955.
65. Frederickson, A.F., Mechanism of weathering. *Geol. Soc. Am. Bull.*, 62, 3, 221–232, 1951.
66. Brown, I.C., Rice, T.D., Byers, H.G., *A study of claypan soils. No. 1488-2016-124425*, United States Department of Agriculture, Washington, D.C. 1933.
67. Larson, W.E., Allaway, W.H., Rhoades, H.F., Characteristics of the Clay Fraction of Various Horizons of Scott Silt Loam and Pawnee Silt Loam 1. *Soil Sci. Soc. Am. J.*, 11, C, 443–447, 1947.
68. Byers, H.G., Alexander, L.T., Holmes, R.S., *The composition and constitution of the colloids of certain of the great groups of soils*, No. 484. US Dept. of Agriculture, United States, 1935.
69. Simonson, R.W., Concept of soil, in: *Advances in agronomy*, vol. 20, pp. 1–47, Academic Press, USA, 1968.
70. Bartelli, L.J., Soil Taxonomy: Its evolution, status, and future, in: *Soil taxonomy—Achievements and challenges*, pp. 7–13, 1984.
71. Richter, D.D. and Babbar, L., II, Soil diversity in the tropics, in: *Advances in ecological research*, vol. 21, pp. 315–389, Academic Press, Cambridge, Massachusetts, 1991.

72. Evans, E.J. and Jackson, M.L., Chemical determination of sorbed water and structural hydroxyl in colloidal minerals of soils and sediments. *Soil Sci. Soc. Am. J.*, 16, 4, 364–368, 1952.
73. Retzer, J.L., Soil Development in the Rocky Mountains. *Soil Sci. Soc. Am. J.*, 13, C, 446–448, 1949.
74. Snider, H.J., The solubility of phosphorus in soils from some Illinois experiment fields. *Soil Sci.*, 38, 6, 471–476, 1934.
75. Nutting, P.G., *Some standard thermal dehydration curves of minerals*, US Government Printing Office, United States, 1943.
76. Nutting, P.G., *The action of some aqueous solutions on clays of the montmorillonite group*, No. 197. US Government Printing Office, United States, 1943.
77. Lange, M.A., Lambert, P., Ahrens, T.J., Shock effects on hydrous minerals and implications for carbonaceous meteorites. *Geochim. Cosmochim. Acta*, 49, 8, 1715–1726, 1985.
78. Grim, R.E. and Rowland, R.A., Differential thermal analysis of clay minerals and other hydrous materials. Part 1. *Am. Min.: Journal of Earth and Planetary Materials*, 27, 11, 746–761, 1942.
79. Norton, F.H., Critical study of the differential thermal method for the identification of the clay minerals. *J. Am. Ceram. Soc.*, 22, 1-12, 54–64, 1939.
80. Grim, R.E., Differential thermal curves of prepared mixtures of clay minerals. *Am. Mineral.: Journal of Earth and Planetary Materials*, 32, 9–10, 493–501, 1947.
81. Graham, E.R., Primary minerals of the silt fraction as contributors to the exchangeable-base level of acid soils. *Soil Sci.*, 49, 4, 277–282, 1940.
82. Graham, E.R., Soil development and plant nutrition: II. Mineralogical and chemical composition of sand and silt separates in relation to the growth and chemical composition of soybeans. *Soil Sci.*, 55, 3, 265, 1943.
83. Graham, E.R., Soil Development and Plant Nutrition. I. Nutrient Delivery to Plants by the Sand and Silt Separates. *Soil Sci. Soc. Am.*, 6, 259, 1941.
84. Reitemeier, R.F., Soil potassium, in: *Advances in agronomy*, vol. 3, pp. 113–164, Academic Press, Cambridge, Massachusetts, 1951.
85. Volk, N.J., The fixation of potash in difficultly available form in soils. *Soil Sci.*, 37, 4, 267–288, 1934.
86. Dennis Rouse, R. and Bertramson, B.R., Potassium Availability in Several Indiana Soils: Its Nature and Methods of Evaluation 1. *Soil Sci. Soc. Am. J.*, 14, C, 113–123, 1950.
87. Jackson, M.L., Hseung, Y., Corey, R.B., Evans, E.J., Vanden Heuvel, R.C., Weathering Sequence of Clay-size Minerals in Soils and Sediments: II. Chemical Weathering of Layer Silicates 1. *Soil Sci. Soc. Am. J.*, 16, 1, 3–6, 1952.
88. Walker, G.F., Trioctahedral minerals in the soil-clays of north-east Scotland. *Mineral. Mag. J. Mineral. Soc.*, 29, 208, 72–84, 1950.
89. Walker, G.F., The decomposition of biotite in the soil. *Mineral. Mag. J. Mineral. Soc.*, 28, 206, 693–703, 1949.

90. Ross, C.S. and Hendricks, S.B., *Minerals of the montmorillonite group: Their origin and relation to soils and clays*, US Government Printing Office, United States, 1945.
91. Foster, M.D., The importance of exchangeable magnesium and cation-exchange capacity in the study of montmorillonitic clays. *Am. Mineral.: Journal of Earth and Planetary Materials*, 36, 9–10, 717–730, 1951.
92. Hendricks, S.B. and Ross, C.S., Chemical composition and genesis of glauconite and celadonite. *Am. Mineral.: Journal of Earth and Planetary Materials*, 26, 12, 683–708, 1941.
93. Kuperman, R.G. and Edwards, C.A., Effects of acidic deposition on soil invertebrates and microorganisms, in: *Reviews of environmental contamination and toxicology*, pp. 35–138, Springer, New York, NY, 1997.
94. Smith, M.C., *Bibliography of the Geology of the Green River Formation, Colorado, Utah, and Wyoming, to July 1, 1986*, Department of the Interior, US Geological Survey, United States, 1986.
95. Chen, D., Lan, Z., Bai, X., Grace, J.B., Bai, Y., Evidence that acidification-induced declines in plant diversity and productivity are mediated by changes in below-ground communities and soil properties in a semi-arid steppe. *J. Ecol.*, 101, 5, 1322–1334, 2013.
96. Jul Mohr, E.C. and Pendleton, R.L., *soils of equatorial regions with special reference to the Netherlands East Indies*, J.W. Edwards, United Nations, 1944.
97. Jul Mohr, E.C. and Pendleton, R.L., The soils of equatorial regions. *LWW*, 58, 5, 410, 1944.
98. Mohr, E.C.J., *The Soils of Equatorial Regions*, Edwards Brothers Inc., Ann Arbor, Michigan, USA, 1944.
99. Fox, R.L., De La Pena, R.S., Gavenda, R.T., Habte, M., Hue, N.V., Ikawa, H., Ikawa, I. *et al.*, Amelioration, revegetation, and subsequent soil formation in denuded bauxitic materials. *Allertonia*, 6, 128–189, 1991.
100. Fox, R.L., de la Pena, R.S., Gavenda, R.T., Habte, M., Hue, N.V., Ikawa, I., Jones, R.C. *et al.*, The revegetation and rehabilitation of strip mined bauxitic soils. National Tropical Botanical Garden, *Allertonia*, 6.2, 128–184, 1991.
101. Sherman, G.D. and Ikawa, H., *Soil sequences in the Hawaiian Islands*, Director of Hawaii Agricultural Experiment Station, Hawaii, 1968.
102. Vageler, P. and Greene, H., *introduction to tropical soils*, Macmillan and Co., Limited, United Nation, 1933.
103. Chenery, E.M. and Hardy, F., The moisture profile in some Trinidad forest and cacao soils. *Trop. Agric.*, 22, 100–115, 1945.
104. Jenny, H., Great soil groups in the equatorial regions of Colombia, South America. *Soil Sci.*, 66, 1, 5–28, 1948.
105. Gupta, I.J. and Crowson, R.A., Zoological Survey of India, in: *State Fauna Series*, vol. 10, pp. 541–57955, 1911.
106. Kanehiro, Y. and Whittig, L.D., *Amorphous mineral colloids of soils of the Pacific region and adjacent areas*, Clay Mineralogy Work Group of the Western Soil and Water Research Committee, Hawaii and California, 1961.

107. Ross, C.S. and Hendricks, S.B., *Minerals of the montmorillonite group: Their origin and relation to soils and clays*, US Government Printing Office, United States, 1945.
108. Thorp, J., Classifying Soils for Mapping in the Rocky Mountain Region 1. *Soil Sci. Soc. Am. J.*, 8, C, 377–378, 1944.

3

Agroecosystems and Bioeconomy

Paolo Di Sia[1,2,3]

[1]*University of Padova, School of Science, Department of Physics and Astronomy & Department of Chemical Science, Padova, Italy*
[2]*University of Padova, School of Medicine, Department of Neuroscience, Padova, Italy*
[3]*Free University of Bozen-Bolzano, Faculty of Science and Technology, Bozen-Bolzano, Italy*

Abstract

The battle against climate change and soil degradation passes also through the land manager function of the farmer. The development of soil health assessment frameworks is important for assessing the effect on soil health of agricultural management systems, evaluating different biological management systems with reduced processing, use of organic fertilizers so as biological control of weeds. The data collection phase relies on the measurement of indicators of bio-chemical-physical and micro-biological health of the soil. The various obtained indices describe the effect of management practices on soil health using a multivariate analysis of the results obtained by indexing techniques. Biological systems have also an increase in biological activity with a better mineralization of carbon and nitrogen, and higher values in enzymatic activities and in the density of earthworms and microarthropods compared to conventional systems. A proposable index is composed of parameters that are most influenced by the organic substance: stability index of the macroaggregates, total nitrogen, SOC, Active-C, mineralizable nitrogen, alkaline phosphatase, ß-glucosidase and leucine aminopeptidase, earthworm density, and QBS-ar. Net benefits involve a greater soil resilience, allowing better adaptation to exceptional events and the attenuation of climate change impact. The bioeconomy of a country includes all the main areas of primary production, i.e., agriculture, forestry, fishing and aquaculture, those of the transformation of biological resources, woodworking, biorefineries, nanobiotechnology, and sea industries.

Email: paolo.disia@gmail.com

Inamuddin, Mohd Imran Ahamed, Rajender Boddula and Tariq Altalhi (eds.) Applied Soil Chemistry, (41–60) © 2021 Scrivener Publishing LLC

Keywords: Agroecosystems, soil chemistry, climate change, soil degradation, soil health assessment frameworks, data collection, nanobiotechnology, bioeconomy

3.1 Introduction

The soil forms the basis of the production of food, fodder, fuel, and fiber, as well as many fundamental ecosystem services. The natural extension of productive soils is limited and must face the increasing pressures of intensification and competing uses of agricultural activities, forestry, pastures, and urbanization, having to satisfy an increasing population which increasingly asks for food, energy, and raw materials. Soils must be recognized and appreciated for their production capacities and for their contribution in guaranteeing food safety and maintaining key ecosystem services.

The soil degradation is due to unsustainable use and management of the territory, but also to extreme climatic events, consequence of various social, economic, and political factors. A part of the world territory is from moderately to highly degraded, due to erosion, salinization, compactation, acidification, and chemical pollution; this rate of degradation threatens the ability of future generations to meet their most essential needs.

There are not many possibilities to expand agricultural areas, except in some parts of Africa and South America. However, many of these lands are not suitable for agriculture and the ecological, social, and economic costs of making them productive would be very high. In world agriculture, the sustainability in managing and producing has therefore become fundamental for reversing the trend of soil degradation and for ensuring current and future world food security [1, 2].

Soil is a resource of fundamental importance but scarcely renewable, which ensures a series of key functions at environmental, social, and economic level. The world population has increased from 2 to 10 million at the dawn of permanent agriculture to more than 7.759 billion at the beginning of 2020. From current demographic trends and expected growth of the world population, which will exceed 9 billion people by 2050, a 60% increase in demand for food, fodder, and fiber is estimated within that year [3, 4].

The environmental problems arising from modern agricultural practices are today associated with radical changes in agroecosystem dynamics, including impacts on plant and animal communities, on the quality of soil, water, and air [5]. The intensification of agriculture in terms of use of water, chemicals, and mechanization, global problems such as climate change, global soil change, and socio-economic problems (demographic boom and globalization of markets) are imposing a growing impact of the human footprint on Earth [6–8].

3.2 Problems Related to Agricultural Intensification

The environmental resource paying the highest price for this growth is soil, from which some major problems related to agricultural intensification have emerged:

1. expansion of soil degradation problems, which highlighted threats such as erosion, decrease in organic matter, local and diffuse contamination, waterproofing, compactation, decline in soil biodiversity, salinization, floods, and landslides [9];
2. negative effects in the displacement of nutrients, together with loss of quality and quantity of support and regulation ecosystem services, whose processes and key properties in the soil are closely connected to the organic substance pool [10, 11];
3. increase in irrigated agricultural areas, consuming more than a quarter of total land evapotranspiration;
4. profound alteration of the global biogeochemical cycle of some greenhouse gases, including carbon (that has undergone a dramatic global depletion in the soils of agroecosystems), nitrogen and methane, with negative impacts on the balance between deposition, fixation in the soil and emission [12, 13];
5. accentuation of the use of fertilizers and pesticides for increasing food production, imposing a stress in terms of soil water retention and water needs of crops following the expected climate change scenarios [14, 15];
6. extinction of many native animal and plant species in agroecosystems, together with impacts on soil biota [16, 17].

If the climate change is having an impact in medium-long term on soil properties, it is verified that the greatest impacts derive from the change in soil use and from its agronomic management; at the same time, the problem becomes particularly critical in front of the future food demand [18, 19].

Evaluating in a holistic way the impacts of agroecosystem management practices, in particular those of the soil, seems therefore necessary to face the future challenges of agricultural sustainability. Soil and crop management is an essential element in the conduct of agroecosystems and in the long-term sustainability of agriculture; this challenge can be addressed through changes in soil quality. The need for a more sustainable soil management has become increasingly urgent as a means of mitigating and adapting to climate change, as well as an important strategy for achieving global food security [20].

Among the various worldwide strategies, an important option is that of carbon sequestration in agroecosystems, especially in agricultural soils; it is a strategy capable of increasing the soil resilience, allowing adaptation to exceptional events and attenuation of impacts of changing rainfall, of increase of temperature and CO_2 levels. The interest of the scientific world tends therefore to shift the attention to sustainable soil and agroecosystem management practices [21, 22].

The research focus is mainly based on the evaluation of the impact that these practices have on the stock and dynamics of carbon in the soil, together with the feedback of the processes related to organic matter in it, and not so much on the impact that carbon undergoes by the climatic change.

There is also criticism about the approach to combat the global food demand through the adoption of a conventional agricultural intensification; the proposal is a compromise and an integration between soil production and conservation (land sharing or wildlife-friendly farming). In the context of environmental science and ecology of agroecosystems, the adoption of a strategy of agroecological intensification of agricultural land is considered more ecoefficient and ecosustainable [23].

Through an efficient ecological-economic compromise, it is demonstrated that the agroecological approach, by promoting the development of biodiversity, is able to increase the quality and quantity of the supply of support and regulation ecosystem services, minimizing environmental costs and maintaining the functional biodiversity of various animal groups [24–26].

The soil should be considered as a real central ecosystem unit within the dynamics of the agroecosystem; in fact, most ecological processes in agroecosystems and natural systems have in the soil their main dynamic regulation center. The use of heavy machinery with deep and continuous plowing and the excessive use of pesticides and fertilizers damage the composition and structure of the soil biological communities and consequently the ecological functions and services provided by the ecosystem "soil" [27].

A holistic approach including the importance of soil biota increases interest in soil biodiversity as an opportunity to study the functional role of biota in improving fertility and for agricultural systems in optimizing the provision of ecosystem services. The biological component of the soil is therefore a key point in the strategy of achieving agricultural sustainability, especially in soil conservation policies and in environmental challenges [28].

The concept of "healthy soil" has emerged in the last decade; it is the soil capable to carrier the production of food and fiber, being able to satisfy human requests, together with a continuous supply of other ecosystem services, necessary for the service of the quality of life and the conservation of biodiversity [29]. The soil is a complex, multicomponent, and multifunctional system, operating at different time and space scales.

In the process of formation of soil aggregates, it is possible to identify as natural capital the mineral components of sand, silt, clay, and the organic substance of the soil in its various components. The size, quantity, and stability of the soil aggregates reflect a balance between aggregate formation factors (soil organic improvers, microfauna, and microflora) and other non-constructive factors (worm bioturbations and deep treatments). There are several functional groups of microorganisms that play a role in the functioning of the soil ecosystem, providing a function by carrying out particular processes.

The functional groups of soil organisms can be defined as a set of species with a similar effect on a specific ecological function. We can consider three functional groups:

- *Chemical regulators*: community of microorganisms that degrade the organic substance to provide nutrients assimilable by plants;
- *Biological regulators*: community of microarthropods, nematodes, and other invertebrates that regulate the population dynamics of other organisms through predation;
- *Ecosystems engineers*: organisms (earthworms, termites, and ants) that modify and create habitats for smaller organisms with the creation of the porosphere and aggregatesphere [30].

The functional groups of organisms are responsible for carrying out some fundamental ecological functions of the soil; the concept of "soil health" derives from the dynamic combination of these functional groups, depending on the chemical and physical conditions of the habitat inside the soil. The choice of different quality indicators for the different ecological functions is an excellent method for evaluating the working of the soil, describing the interplays taking place between processes and diverse components of soil health.

The concept of soil health, based on the ecosystem processes carried out by the organisms constituting the soil biodiversity, is considered as an integrative property of each agroecosystem, including the component influenced by soil management [31]. The quality, biodiversity, and resilience of the soil are extremely sensitive to anthropogenic disturbance. Soil health can therefore be considered the main link between agronomic management practices and the achievement of the major objectives of sustainable agriculture.

The objectives of managing an agroecosystem are therefore to coordinate a stable productivity with the sustainability of the ecological environment, implementing a health state of the soil; this means to know how to manage matter and energy in order to favor the supply of the necessary ecosystem services for the sustainable use of resources. The challenge is therefore to be able to combine an ecoefficient soil management, increasing the soil health through a positive balance of carbon and nutrients, with the conservation of functional biodiversity and including the part of the soil biota as an essential component of its integrated management within the dynamics of the agroecosystem [32–34].

3.3 Important Themes and Indicators

The soil quality literature generally focuses on three major themes:

- *education,*
- *evaluation,*
- *development of kit tests.*

Regarding the evaluation, the effective features consist of:

- a Minimum Data Set (MDS) of indicators related to soil functions with a strong influence in assessing the effect of management practices and soil degradation;
- a MDS of soil health indicators with the following characteristics:
    - sensitive to soil management practices;
    - representative of soil ecological functions and well correlated with ecosystem processes;

- to have quantifiable threshold values, in order to be able to evaluate the effects of management practices;
- to be insertable and manageable within a methodology providing information to allow to consistently evaluate the positive or negative effect of the adopted and/or to be adopted practices.

Other characteristics of good indicators concern the ability to integrate biological, chemical, and physical properties of processes and functions difficult to measure directly, in an inexpensive way in terms of time and money. Among the various proposed indicators we remember the "Cornell Soil Health Assessment" (CSHA) [35] and the "Soil Management Assessement Framework" (SMAF) [36]. The latter, in particular, evolved from studies that applied the principles of natural systems engineering and soil ecology to interpret holistically the physical, chemical, microbiological, and biological data collected from various studies on soil management.

The index creation phases are essentially based on:

1. choice of indicators;
2. transformation of indicators into scores;
3. integration of indicators into an SHI (Soil Health Index).

The data combination methodology is the one used in most evaluation studies. The ecological functions initially proposed and subsequently improved concern:

- *Nutrient cycling*: soils working well are those having a high potential in providing optimal quantities of main nutrients and immobilize nutrients that can be dangerous or harmful to plants if released into air/water;
- *Water movement and availability*: the flow of free water is significant to provide water in the root zone of the plant and to consent the movement in solution of nutrients and bounteous organisms of the soil. The partitioning and conservation of water and solutions can help in maximizing the deep percolation in the recharge of aquifers and in helping the soil to resist to erosive forces;
- *Physical firmness and structural support*: soils working well have a physical structure providing a means for the growth of the root system of plants and for resisting to the caustic forces of wind and water. The soil structure is tightly related

and often indispensable for many other functions. In addition, stable aggregates take and stabilize carbon more efficiently and in the long term;
- *Filtering and buffering*: soils possesses a natural ability to degrade caustic or dangerous compounds. When they work suitably, can treat modest amounts of present contaminants, making them less dangerous to plants and animals, often deteriorating the compound or absorbing a certain amount;
- *Biodiversity and habitat*: it relates to the natural ability of the soil to supply the necessary conditions to carrier a variety of healthy plants and animals. From an agronomic point of view, it is significant for the integrated management of pathogens, for the nutrient cycle and for the health of the agroecosystem.

3.4 The Bioeconomy

"Bioeconomy" is the socio-economic system including and interconnecting the economic activities that habit renewable bio-resources of soil and sea, such as agricultural crops, forests, animals, terrestrial and marine micro-organisms, to build food, materials, and energy.

Parts of bioeconomy are the various sectors of primary production, i.e., agriculture, forests, fishing, aquaculture, the industrial sectors using or transforming the bio-resources coming from previous sectors, such as the food industry, the cellulose, and paper industry, part of the energy chemical and biotechnological industry [37–41]. Bioeconomy represents the heart of sustainable development, a virtuous path that leads communities toward an environmentally friendly economy, in which the dependence on fossil fuels and non-renewable resources is reduced.

In this way, the great transformations in the soil use are limited, regenerating the environment, creating new economic increase and employment, starting from local specificities and traditions in rural, coastal, and industrial areas. The global demographic increase, the climate change, and the reduction of the resilience capacity of ecosystems require an increase in the use of renewable biological resources for a more sustainable primary production, so as more capable transformation systems for food, fibers, and other bio-based products production, with less production of waste gas emissions, and consequent benefits for human health and environment.

The enhancement of organic waste from agriculture, forests, cities, and industry completes the action by guaranteeing the bioeconomy a key role in the circular economy [42]. This strategy aims to supplement the sustainable production of renewable biological resources and their conversion into a variety of value-added products; it also has the purpose to offer a shared vision on economic and environmental opportunities and on challenges connected with the implementation of the bioeconomy rooted in territories.

The actions to determine the achievement of the indicated objectives can be summarized in the following points:

- *to improve the sustainable and quality production of products* in each of the indicated sectors (from those of primary production to those of transformation), exploiting their interconnections more efficiently, with a punctual enhancement of both land and marine biodiversity, of ecosystem services and of circularity;
- *to create new "value" chains*, longer and more deeply rooted in territories, to allow the regeneration of abandoned areas, marginal lands and industrial sites;
- *to create greater investments in research and innovation*, spin offs and start-ups, education, training, and communication;
- *to improve the coordination* between stakeholders and regional, national, and community level policies;
- *to improve people involvement*;
- *to conduct targeted actions* for market development [43].

3.5 On Circular Bioeconomy

Food security, sustainable administration, and exploitation of agricultural soils, forests, marine flora, fauna and water are among the circumstances that most influence the society worldwide. Circular bioeconomy works in particular within four macro-sectors:

1. *Agri-food*;
2. *Forests*;
3. *Bioindustry*;
4. *Marine bioeconomy*.

Thinking about a transition to the circular economy means considering the bioeconomy within a vision in which the production and use of renewable bioresources and their conversion into products with high added value is part of a circular system, making economic activities more profitable and sustainable in the long time. A good circular bioeconomy needs to be built both on local resorts and infrastructure and through and improved interaction and integration between economic areas, public and private stakeholders, and existing policies at national and international level [44].

In Europe, bioeconomy has reached a value of more than 2,000 billion euros in annual turnover, with more than 20 million jobs, and an increase in market value of additional billion euros and tens of thousands new jobs is expected in 2020 and beyond [45, 46].

The food industry is the first in size within the EU and is potentially expanding, with new markets and industries arising in the food and non-food sectors, both new and traditional. With a sustainable management of water and food chain, it is possible to contribute to solve the biggest problems of nutrition, health, and social well-being. Local biodiversity, water resources, tourism, energy production from renewable sources, marine aquaculture represent important local chances for "blue" growth and employment, still not expeditiously exploited [47].

3.6 Bioeconomy, Environment, and Natural Capital

From an environmental point of view, bioeconomy brings chances and challenges. Chances are connected to the gradual transition, within the production processes, from the use of non-renewable resources to renewable ones, to bound the environmental stress on ecosystems and enhance the value for their conservation, as a source of relevant services also for the economy. The enhancement of production activities deriving by renewable sources can facilitate the management of waste, as these sources are more reassimilable.

We have several examples of unsustainable conduct for the environment and human health, especially in the food and fish industry sector. Often, there is no necessity to increase the production of raw materials, but rather to increase their added value and usefulness ameliorating the quality of products. It is notable to ensure the link between economic development and environmental necessity; the activity of the bioeconomic sectors based on the import of renewable raw materials from countries with less stringent environmental regulations should be avoided, as this favors the transfer of environmental damage to foreign countries. The principles to follow should be *to make better use of what we yet use* and *to use effectively what we do not yet use*. An agriculture based on sustainable training responds to its primary function of food production, simultaneously supporting the correct conduct of natural resources (soil, water, air, and biodiversity) and contributes to the conservation of genetic diversity.

Nature offers a wide range of crude materials from which it is possible to synthesize chemical intermediates similar to those obtained from fossil ones, as well as a wide kind of molecules and interesting unexplored synthesis processes. Vegetable oils, corn and potato starch, cellulose extracted from straw and wood, lignin and aminoacids, and the exploitation of algae are becoming increasingly important industrial raw materials.

Using bio-chemical-physical processes, these materials can be changed into fuel, chemical intermediates, polymers, and other macromolecules, for which mineral oils have been used so far. In terms of use of raw materials, within an innovative and sustainable bioeconomy, more and more industries should use the following lines:

- *identification of species* (identification of indigenous crops, plants, and algae) *and of the most suitable biomass* (devoted low-impact crops or non-fragile marine areas) to be used in biorefinery processes, limiting the emissions of climate-changing gases associated with the land use, with sharing of the benefits deriving by the use of local genetic resorts;
- *identification and use of economically peripheral and/or contaminated land*, preventing the excessive use of soil and native landscapes, in line with international policies on climate exchange and biodiversity;
- *promotion of the competent use of biomass*, extracting any possible value from renewable sources, starting by food and feed components;
- *construction of new production chains* with the agricultural and forestry world;
- *promotion of forest products* with certified ecolabels, preferably by local supply chains.

The use of biowaste for producing materials and energy is the basis of a maintainable bioeconomy that reduces the pressure on primary renewable sources and decreases the level of dependency on fossil fuels. Organic waste includes those deriving from primary production of vegetable origin (straw from cereals, tops, and sugar cane leaves) and animal origin, together with subproducts and waste deriving by the primary transformation of biomass of food origin and by food processes (peels, shells, and bagasse).

There are also residues of products of forestry origin (tops and branches) and residues of primary processes deriving from sawmills or paper mills (sawdust, shavings, and lye/black liquor), as well as the organic fraction of produced and collected urban waste. These waste streams are poorly valued and mostly eliminated in landfills, with significant costs and unfavorable environmental impacts. Many of them can be a substantial source of chemicals and bio-based materials, together with substrates for biotechnological production that can provide value-added products, such as fine chemicals, bio-based materials, and fuels [48, 49].

3.7 The Bioeconomy in Daily Life

There are interesting examples of bioeconomy present in daily life:

- *Biodegradable shopping bags*: with their double use (bags for purchases and then bags for waste), they facilitate the separate harvesting of organic waste and reduce the risks of dispersion in the environment. Being biodegradable, compostable, transparent, and resistent to humidity and pathogens, they are a key tool to favor the separate collection of organic waste and for the production of high quality compost.
- *Biogas*: it is an organic gas made by the anaerobic digestion of biomass, useful as a fuel for the fruition of electricity and heat.
- *Catering items*: the use of biodegradable and compostable items for catering makes simpler post-consumer disposal operations. These items can be placed with the organic fraction in composting facilities, thus reducing the impact of waste on the environment and the related management costs.
- *Packaging items*: these items can be disposed with organic waste. Interesting applications are food paper and compostable trays, extruded and interwoven nets, toilet paper, napkins, magazine films, and compostable fruit labels.
- *New cold-pressed oils*: these are innovative vegetable oils based on olives and tomatoes, extracted solely through mechanical processes. They are a valid alternative to the chemical extraction of functional compounds and are appropriate for vegetarians, vegans and gluten-free diets.

- *Envelopes and gloves for fruit and vegetables*: the transition from normal plastic products to biodegradable and compostable ones helps to decrease the production of not easily recyclable waste and to rise the collection of organic waste.
- *Compostable capsules*: compostable capsules can be gathered with organic waste and sent for the industrial composting process, in which the used capsule and coffee are recuperated together in compost, with an important reduction in the volume of waste and emissions of greenhouse gases.
- *Carboxylic and cosmetic acids*: carboxylic acids are basic ingredients of plastic materials of biological origin, and intermediates in the creation of plasticizers, for complex esters used in the sector of lubricants, pharmacosmetic formulations, and herbicides. Several carboxylic acids with their esters and glycerin are products generally used in cosmetic applications.
- *Lubricants*: they are a perfect solution for machinery active in ecologically sensitive areas (agricultural, forestry, marine, and urban areas), providing advantages in terms of both performance and environmental impact and safety.
- *Bioactive compounds extracted from meat subproducts*: this is an optimization and validation of the modified response surface methodology for the extraction of bioactive peptides from fish and poultry subproducts.
- *Plasticizers*: these are plasticizers for special elastomers and PVC, so as a new type of bio-extender oil for elastomers for general use.
- *Digested pellet*: it is an organic fertilizer full of nutrients, elements such as nitrogen, phosphorus, and potassium, created by anaerobic digestion of agricultural biomass.
- *Bio-innovation in recipes*: usage of natural products to replace those derived by chemical synthesis processes, to carrier the diversification of products and diets, for example for snacks and desserts using derivatives of spirulina seaweed, full of essential nutrients such as proteins, vitamins, minerals, aminoacids, and Omega 3 and 6.
- *Patch with low environmental impact*: it consists of a support composed of a film based on bioresins derived from potato starch (100% biocomposable), with a central pad without preservatives. It is an innovative glue without natural rubber latex and without solvents, in a package made of recycled cardboard.

- *Bioethanol*: it is a renewable energy source deriving from the fermentation of the sugar and starch components of the plant subproducts, mainly of sugar cane.
- *Mulching film*: the organic and biodegradable film for mulching is an efficient alternative from an agronomic and environmental point of view compared to traditional mulching film. It decomposes in the soil so as not to require removal at the end of the season and allows efficient control of weeds and equivalent agronomic harvests, in terms of both quantity and quality [50–54].

3.8 Conclusions

The carried-out processes allow, together with the contribution of organic fertilizers, a more stable structure. If, on one hand, a greater stability allows to resist the driving force of heavy rains, on the other one, it leads to a stabilization of carbon in the soil. About the long-term stabilization of carbon in agroecosystems, the activity of earthworms seems to be decisive, which together with microartropods present well-structured communities in abundance compared to conventional systems.

The dynamics of carbon and nitrogen mineralization in biological systems are governed by a greater proximity to a metabolic balance typical of mature agroecosystems. Confirming the low biological activity of conventional systems, the enzymatic activities testify for these systems a poor ability to degrade some of the most important chemical bonds present in organic compounds. The most significant differences between biological and conventional systems are found for the activities of ß-glucosidase, leucine aminopeptidase, and arylsufatase.

In support of the experimental observations, interesting are the evaluation approaches based on the ecological services and functions provided by the soil to examine the links between soil health, soil management practices, and endpoints within the agroecosystems, evaluating so the soil health indices. The number of annual workings, the application of pesticides, and the percentage of cover crops are the indisputable endpoints to be represented by soil health indices as important indicators of sustainability of agroecosystems.

The results of the transformation of indicators into scores, attributable to the expressed level of the ecological function, are the crucial phase in the creation of the index. Some indicators have the best characteristics for the construction of a simpler, faster, useful, and representative index of soil health in the agroecosystem. A proposable index is composed of the parameters most influenced by the organic substance:

- for physical component: macroaggregate stability index, average aggregate diameter, and apparent density;
- for chemical component: total nitrogen;
- for bio-chemical and microbiological component: Active-C, nitrogen (PMN), alkaline phosphatase, ß-glucosidase, and leucine aminopeptidase;
- for bio-diversity: earthworm density and QBS-ar.

A carefully chosen set of indicators like these ones well represents the components of soil health; it can adequately provide the necessary information to decision makers.

The methodology is flexible and can therefore undergo changes to obtain transformation algorithms and integrative capacity of the final indices, capable of describing the effect of management practices on a large set of ecological functions and in different types of agroecosystems.

The use of assessment tools therefore effectively reflects management and sustainability objectives; they can be used to promote sustainability in decisions regarding the management of agroecosystems in rural development plans, soil conservation policies, green credits, and other applications.

In order for the bioeconomy to overcome the defiance of reintegrating economy, society and environment, it is not enough to use biomass for industrial applications and to practice renewable raw materials instead of fossil ones. In fact, it is not only a matter of integrating biological knowledge into existing technology, but the transition must also take place at a social level, stimulating awareness and dialog, and further supporting innovation in social structures by promoting more aware behaviors.

A better knowledge of what is consumed, in particular of food products and related processes, will promote the improvement of people's health and lifestyle, pushing for sustainable innovation by businesses. This transition process in economy and society requires a holistic approach, with the citizens being the real protagonists of the social transformation that the bioeconomy can produce.

Bioeconomy is also a challenge for reconnecting environment, economy, and society, generating economic value together with new values and a new cultural approach. The challenge requires the following:

- business models by companies that involve customers, workers, and citizens, in a common vision of sustainability; this creates new economic value, employment, relationships, and interactions that are able to meet social needs and improve the well-being of community;

- at the public level, the widespread adoption of both a participatory local approach and a new idea of territory, understood as a localized set of tangible and intangible assets and relations between public and private entities. Being aware of the territorial distribution of renewable resources, strengths and weaknesses, needs and barriers to development, allows to recompose fragmented skills and knowledge into new flows of productive consciousness, forming an innovation matrix and helping to create a new territorial identity;
- to recognize the growing need to preserve soils and ensure that governments make the right investments;
- to promote management practices for adaptation to climate change and the mitigation of its effects, for resilience to evolving climate models and extreme weather conditions;
- to promote strict regulations and effective controls by governments to limit the accumulation of contaminants beyond the pre-established thresholds for human health;
- to strengthen the implementation of educational programs on sustainable soil management.

References

1. Fao, Agroecology, http://www.fao.org/agroecology/home/en/, 2020.
2. Zhu, W., Wang, S., Caldwell, C.D., Pathways of assessing agroecosystem health and agroecosystem management. *Acta Ecol. Sin.*, 32, 9, 2012.
3. Karlen, D.L., Ditzler, C.A., Andrews, S.S., Soil quality: why and how? *Geoderma*, 114, 145, 2003.
4. Worldometers, https://www.worldometers.info/it/, 2020.
5. Stoate, C., Báldi, A., Beja, P., Boatman, N.D., Herzon, I., Van Doorn, A., De Snoo, G.R., Rakosy, L.E., Ramwell, C., Ecological impacts of early 21st century agricultural change in Europe - A review. *J. Environ. Manage.*, 91, 1, 22, 2009.
6. Pretty, J., Agricultural sustainability: concepts, principles and evidence. *Philos. Trans. Royal Soc. B*, 363, 1491, 447, 2007.
7. Kitzes, J., Wackernagel, M., Loh, J., Peller, A., Goldfinger, S., Cheng, D., Tea, K., Shrink and share: humanity's present and future ecological footprint. *Philos. Trans. Royal Soc. B*, 363, 467, 2008.

8. Richter, D.B., Humanity's transformation of earth's soils: pedology's new frontier. *Soil Sci.*, 172, 12, 957, 2007.
9. Ruttan, V.W. (Ed.), *Sustainable Agriculture and the Environment - Perspectives on Growth and Constraints*, Routledge, London and New York, 2019.
10. Lal, R., Anthropogenic Influences on World Soils and Implications to Global Food Security. *Adv. Agron.*, 93, 69, 2007.
11. Tilman, D., Cassman, K.G., Matson, P.A., Naylor, R., Polasky, S., Agricultural sustainability and intensive production practices. *Nature*, 418, 8, 2002.
12. Savary, S., Ficke, A., Aubertot, J., Hollier, C., Crop losses due to diseases and their implications for global food production losses and food security. *Food Secur.*, 4, 519, https://doi.org/10.1007/s12571-012-0200-5, 2012.
13. Galloway, J.N., Dentener, F.J., Capone, D.G., Boyer, E.W., Howarth, R.W., Seitzinger, S.P., Asner, G.P., Cleveland, C.C., Green, P.A., Holland, E.A., Karl, D.M., Michaels, A.F., Porter, J.H., Townsend, A.R., Nitrogen cycles: past, present, and future. *Biogeochemistry*, 70, 153, 2004.
14. Impact of Climate Change on Biodiversity, http://www.naro.affrc.go.jp/archive/niaes/marco/marco2015/text/plenary-2_b_marambe.pdf, 2020.
15. Olesen, J.E., Trnkab, M., Kersebaumc, K.C., Skjelvagd, A.O., Seguine, B., Peltonen-Sainiof, P., Rossig, F., Kozyrah, J., Micalei, J., Impacts and adaptation of European crop production systems to climate change. *Eur. J. Agron.*, 34, 96, 2011.
16. Brussaard, L., De ruiter, P.C., Brown, G.G., Soil biodiversity for agricultural sustainability. *Agric. Ecosyst. Environ.*, 121, 233, 2007.
17. Firbank, L.G., Petit, S., Smart, S., Blain, A., Fuller, R.J., Assessing the impacts of agricultural intensification on biodiversity: a British perspective. *Philos. Trans. Royal Soc. B*, 363, 1492, 777, 2008.
18. Hazell, P. and Wood, S., Drivers of change in global agriculture. *Philos. Trans. Royal Soc. B*, 363, 467, 2008.
19. Godfray, H.C.J., Beddington, J.R., Crute, J.I., Haddad, L., Lawrence, D., Muir, J.F., Pretty, J., Robinson, S., Thomas, S., Toulmin, C., Food security: the challenge of feeding 9 billion people. *Science*, 327, 812, 2010.
20. Benton, T.G., Vickery, J.A., Wilson, J.D., Farmland biodiversity: is habitat heterogeneity the key? *Trends Ecol. Evol.*, 18, 182, 2003.
21. Lal, R., Sequestering carbon in soils of agro-ecosystems. *Food Policy*, 36, S33, 2011.
22. Jacobson, M.Z., Review of solutions to global warming, air pollution, and energy security. *Energy Environ. Sci.*, 2, 148, 2009.
23. Vaarst, M., Escudero, A.G., Chappell, M.J., Brinkley, C., Nijbroek, R., Arraes, N.A.M., Andreasen, L., Gattinger, A., De Almeida, G.F., Bossio, D., Halberg, N., Exploring the concept of agroecological food systems in a city-region context. *Agroecol. Sustainable Food Syst.*, 42, 6, 686, 2018.
24. Lescourret, F., Magda, D., Richard, G., Adam-Blondon, A.-F., Bardy, M., Baudry, J., Doussan, I., Dumont, B., Lefèvre, F., Litrico, I., Martin-Clouaire, R., Montuelle, B., Pellerin, S., Plantegenest, M., Tancoigne, E., Thomas, A.,

Guyomard, H., Soussana, J.-F., A social-ecological approach to managing multiple agro-ecosystem services. *Curr. Opin. Environ. Sustain.*, 14, 68, https://doi.org/10.1016/j.cosust.2015.04.001, 2015.
25. Dominati, E., Patterson, M., Mackay, A., A framework for classifying and quantifying the natural capital and ecosystem services of soils. *Ecol. Econ.*, 69, 1858, 2010.
26. Kibblewhite, M.G., Ritz, K., Swift, M.J., Soil health in agricultural systems. *Philos. Trans. Royal Soc. B*, 363, 685, 2008.
27. Lavelle, P., Decaëns, T., Aubert, M., Barot, S., Blouin, M., Bureau, F., Margerie, F., Mora, P., Rossi, J.P., Soil invertebrates and ecosystem services. *Eur. J. Soil Biol.*, 42, 1, 3, 2006.
28. Doran, J.W. and Zeiss, M.R., Soil health and sustainability: managing the biotic component of soil quality. *Appl. Soil Ecol.*, 15, 3, 2000.
29. Al-Kaisi, M.M. and Lowery, B. (Eds.), *Soil Health and Intensification of Agroecosytems*, Academic Press, Cambridge, MA, 2017.
30. Fortuna, A., The Soil Biota. *Nature Education Knowledge*, 3, 10, 1, 2012.
31. Herrick, J.E., Soil quality: an indicator of sustainable land management? *Appl. Soil Ecol.*, 15, 75, 2000.
32. Kummu, M., de Moel, H., Porkka, M., Siebert, S., Varis, O., Ward, P.J., Lost food, wasted resources: Global food supply chain losses and their impacts on freshwater, cropland, and fertiliser use. *Sci. Total Environ.*, 438, 477, https://doi.org/10.1016/j.scitotenv.2012.08.092, 2012.
33. Drinkwater, L.E. and Snapp, S.S., Nutrients in Agroecosystems: Rethinking the Management Paradigm. *Adv. Agron.*, 92, 163, 2007.
34. Anderson, T.H. and Domsch, K.H., The metabolic quotient for CO_2 (qCO_2) as a specific activity parameter to assess the effect of environmental conditions, such as pH, on the microbial biomass of forest soils. *Soil Biol. Biochem.*, 25, 3, 393, 1993.
35. Soilhealth, Cornell, https://soilhealth.cals.cornell.edu/, 2020.
36. Wienhold, B., Karlen, D., Andrews, S., Stott, D., Protocol for indicator scoring in the soil management assessment framework (SMAF). *Renewable Agric. Food Syst.*, 24, 4, 260, 2009.
37. Di Sia, P., Present and Future of Nanotechnologies: Peculiarities, Phenomenology, Theoretical Modelling, Perspectives. *Rev. Theor. Sci.*, 2, 2, 146, https://doi.org/10.1166/rits.2014.1019, 2014.
38. Di Sia, P., Nanobiomaterials for environmental protection: state of the art, applications and modelling. *Int. J. Eng. Innov. Res. (IJEIR)*, 3, 5, 688, 2014.
39. Di Sia, P., Present and Future of Nano-Bio-Technology: Innovation, Evolution of Science, Social Impact. *Online J. Educational Tech. (TOJET), Special Issue 2 for INTE 2015*, 442, 2015.
40. Di Sia, P., Nanotechnologies among Innovation, Health and Risks. *Procedia – Soc. Behav. Sci. J.*, 237, 1076, http://dx.doi.org/10.1016/j.sbspro.2017.02.158, 2017.
41. Di Sia, P., Nanotoxicology and human health. *World Sci. News*, 100, 86, 2018.

42. Geissdoerfer, M., Savaget, P., Bocken, N.M.P., Hultink, E.J., The Circular Economy - A new sustainability paradigm? *J. Cleaner Prod.*, 143, 757, https://doi.org/10.1016/j.jclepro.2016.12.048, 2017.
43. Elia, V., Gnoni, M.G., Tornese, F., Measuring circular economy strategies through index methods: A critical analysis. *J. Cleaner Prod.*, 142, 4, 2741, https://doi.org/10.1016/j.jclepro.2016.10.196, 2017.
44. El-Chichakli, B., von Braun, J., Lang, C., Barben, D., Philp, J., Policy: Five cornerstones of a global bioeconomy. *Nature*, 535, 7611, 221, 2016.
45. Ec Europa, Research, Bioeconomy, https://ec.europa.eu/research/bioeconomy/pdf/official-strategy_en.pdf, 2020.
46. Ec Europa, Research, Bioeconomy, Strategy, https://ec.europa.eu/research/bioeconomy/pdf/ec_bioeconomy_strategy_2018.pdf, 2020.
47. Eumofa, Blue Bioeconomy, https://www.eumofa.eu/documents/20178/84590/Blue+bioeconomy_Final.pdf, 2020.
48. Bennett, L.T., Mele, P.M., Annett, G.S., Kasel, S., Examining links between soil management, soil health, and public benefits in agricultural landscapes: An Australian perspective. *Agric. Ecosyst. Environ.*, 139, 1, 2010.
49. McCollum, D.L., Krey, V., Riahi, K., Kolp, P., Grubler, A., Makowski, M., Nakicenovic, N., Climate policies can help resolve energy security and air pollution challenges. *Clim. Change*, 119, 479, https://doi.org/10.1007/s10584-013-0710-y, 2013.
50. Di Sia, P., Agri-food sector, biological systems and nanomaterials, in: *Food Applications of Nanotechnology*, 1st ed, G. Molina, Inamuddin, F.M. Pelissari, A.M. Asiri, (Eds.), CRC Press, Boca Raton, 2019.
51. Power, A.G., Ecosystem services and agriculture: tradeoffs and synergies. *Philos. Trans. Royal Soc. B*, 365, 2959, 2010.
52. Materbi, Agriculture, Mulching Film, http://materbi.com/en/solutions/agriculture/mulching-film/, 2020.
53. Di Sia, P., The Nanotechnologies World: Introduction, Applications and Modeling, in: *Fundamentals and Applications*, vol. 1, S. Sinha, and N.K. Navani, (Eds.), Studium Press, Houston, 2013.
54. Di Sia, P., *Mathematics and Physics for Nanotechnology - Technical Tools and Modelling*, Jenny Stanford Publishing, CRC Press, Boca Raton, 2019.

4

Technological Advances in Analyzing of Soil Chemistry

M. Ramesh[1]* and L. Rajeshkumar[2]

[1]Department of Mechanical Engineering, KIT-Kalaignarkarunanidhi Institute of Technology, Coimbatore, Tamil Nadu, India
[2]Department of Mechanical Engineering, KPR Institute of Engineering and Technology, Coimbatore, Tamil Nadu, India

Abstract
Soil plays a vital role in nurturing all the living creatures in the world and its nature needs to be effectively maintained. The nature of soil changes owing to the variation in the chemical composition that are present in the soil. The chemical composition constantly varies because of many factors like weather and vegetation, so monitoring soil chemical composition is one of the most important factors. The parameters relating to chemical composition of soil are the pH value, organic substances in soil, kind of nutrients it possess, and ion capacity for exchanging. Digital technology, such as sensors, the Internet of Things (IoT), machine learning (ML), artificial intelligence (AI), big data, as well as the use drones and GPS satellites, are now increasingly employed in agriculture, crop science and soil analysis. This chapter discusses all these technologies and how they relate to soil chemistry and composition.

Keywords: Soil chemistry, soil temperature, soil moisture, soil humidity, organic compounds, artificial intelligence, big data, Internet of Things, machine learning

4.1 Introduction

Soil is the most essential part of the ecosystem and for all lives. It accommodates various lifeforms such as plants and animals, maintains the

**Corresponding author*: mramesh97@gmail.com

Inamuddin, Mohd Imran Ahamed, Rajender Boddula and Tariq Altalhi (eds.) Applied Soil Chemistry, (61–78) © 2021 Scrivener Publishing LLC

atmospheric gas pressures, and filters the surface water. Agriculture depends purely on the soil and crop yield is dependent on the quality of soil. Soil is responsible for spreading the water and fertilizer to the crops and its roots uniformly. Parameters like soil temperature, electrical conductivity, soil moisture, soil pH, organic compounds, and soil humidity govern the chemical and physical properties of the soil and these parameters has to be continuously observed by the farmers to have better control over it [1]. Soil chemistry analysis is becoming so important these days as it is considered to be a field of science having much more scope for research and explorations. Most of the environmental hazards including food spoliation, change in climatic condition, ground water depletion, and contamination are directly connected with soil. Some explicit reactions due to change in soil composition are also occurring every now and then at various locations. A soil analyst primarily took this field for research and gave out major findings [2]. Yet, there are a lot more to explore in this field including the ion exchange behavior of soils. Such works require interdisciplinary collaborations and are awaiting cutting edge discoveries.

4.2 Soil Chemistry

Molecular and nano-sized mechanisms can easily be described by many recently invented contemporary investigative methods such as synchrotron-based techniques as far as soil chemistry concerned. Many researchers elucidated synchrotron-based techniques for their data analysis, methodology, and preparation of samples [3–6] and vibrational spectroscopy [7]. Adsorption of a compound or complex depends upon many factors of environment such as time of sorption, sorbent type, surface loading, pH value at the time of sorption, and ionic strength. Soils usually and mostly contain adsorption complex continuum within them [8–11]. Hence, there could be a continuum present amid surface precipitation and adsorption. At low surface loads, adsorption of complexes either in the outer or inner sphere dominates, while at medium surface loadings, aggregates of compounds are formed on the soil surfaces, and at high surface loadings, precipitation of surfaces tends to be predominant.

Exterior complex sphere of the soil molecules comprises of earth alkaline cations like Ba^{2+}, Mg^{2+}, Sr^{2+}, and Ca^{2+} which were determined by conventional macroscopic analysis techniques. Meanwhile the inner complex sphere of the soil are made up of divalent metal heavy cations such as Pb^{2+}, Cd^{2+}, and Hg^{2+} along with transition metal cations such as Fe^{2+}, Zn^{2+}, Mn^{2+}, Ni^{2+}, Cu^{2+}, and Co^{2+} which were determined by molecular level analysis methods like

X-ray absorption spectroscopy (XAS). Precipitation occurs at the surface of the metal oxides and phyllosilicates that adsorb metallic elements like zinc, nickel, chromium, and cobalt during higher loads and pH values [12–16]. Surface complex elements formed by the adsorption of compounds like NO^{3-} on the surface of soil minerals and these are formed on the surfaces with positive charge and on the outer sphere of the complex elements. Yet, the determination methods of the type of complex could not be carried out due to the absence of molecular level data of the soil complex. Specific compounds like SO_4^{2-} are adsorbed as outer sphere element which was proven by few researchers but some characterization techniques like spectroscopic methods revealed that the elements were also adsorbed as inner sphere complexes [17–19]. Meanwhile, this controversy is further supported by various other environmental conditions like pH of the soil, ionic strength, and concentration, which has given different inferences. Some researchers has given a conclusion that selenite compounds were adsorbed as inner and outer sphere complexes based on the environmental parameters [2, 20, 21]. Some other anionic elements like phosphate, molybdate, silicate, arsenate, and arsenide are strongly adsorbed only in the inner sphere complexes through ligand exchange contrivances but these adsorptions were not mostly dictated by ion exchange strengths. Ligand exchange mechanisms always result in change of pH value from most alkaline to most acidic [22, 23].

According to some researchers soil elements like clays, metal oxides, and phyllosilicates could act as substrate for the formation of precipitates and multi-nuclear complexes of Ni^{2+}, Co^{2+}, Cu^{2+}, Cr^{3+}, and Pb^{2+}. When the thermodynamic solubility of the products is concerned, the above said metal oxides are formed at surface loadings well beneath the monolayer coverage and in pH ranges below which the precipitation of the metal oxides occur [24–27]. Some researchers proved that, apart from precipitation, surface co precipitation could also occur due to the mixed metal hydroxide formation when metals like nickel were adsorbed on aluminium oxide and phyllosilicates [28]. Structural commonness could be observed between such precipitates when compared with layered double hydroxides (LDH) and hydrotalcite mineral groups. Structure of typical LDH is shown in Figure 4.1. Metal sequencing is easily carried out by structures arranged by LDHs. Particularly, metals like zinc, nickel, and cobalt were constantly released by the soil at regular intervals in spite of age of soil precipitates and prevent these metals subjected to leaching and being washed away by microbes and plants intake. Enhanced silication of LDH interlayer along with dwell time reduces the bio-availability and release of metals thus rendering the formation of phyllosilicate precipitate precursor from LDH [29–31].

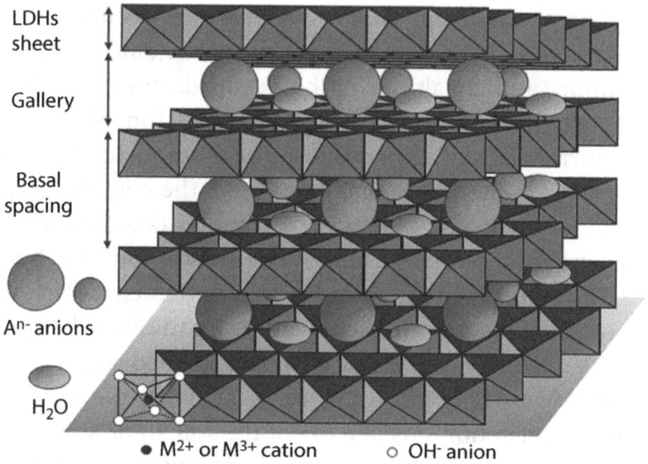

Figure 4.1 Structure of LDH [30].

Researches on the kinetics of LDH and its related details were very meagerly available till date. This limited study is due to the fact that the precipitation process on the metal surfaces usually took loner times. On contrary, Siebecker *et al.* experimented the precipitation time of Ni-Al LDH phases on pyrophillite and concluded from Q-XAS results that the time was as minimum as 6 to 15 min. Reports from the authors further revealed that precipitation and adsorption may occur simultaneously with almost equal time spans. From this, the necessity of strong model of adsorption and the presence of continuum could be evidently witnessed which has a capability of recording many concurrent adsorption phenomenon within the time of reaction. Field soils also contain LDH phases where the metals are impounded within them in high quantity. This metal impounding is thus observed at high pH values, and loadings over metals so that metals like zinc, nickel, and cobalt remain the impure soils. Their bio-availability in the soils remain enhanced and the mobility is also encumbered [32].

4.2.1 Advances in Time-Resolved Molecular Scale Techniques

The chemical reaction at the interface between minerals and water in soil conditions may occur at a time scale ranging between microseconds and years. Numerous salient reactions including reduction, adsorption, precipitation, and oxidation were considered to occur initially by a time frame ranging from milliseconds to minutes at the surface of the minerals in the soils. It was also said that most of the reactions on the mineral surfaces terminates within the above said time well before the commencement of nuptial measurement of process parameters using flow and conventional methods. In order to understand the environmental chemical processes, finding of chemical kinetic rate constants and mechanisms is very important while the data regarding secondary reactions and back reactions could not be obtained for such mechanisms whose reaction time is very minimum [33]. For recording data in milliseconds time chemical relaxation methods like concentration jump and pressure jump methods are used for swift recording of data. However, in such studies the chemical rate constants are not determined directly but obtained by using linear rate equations that are rendered from modelling analysis and equilibrium conditions.

Measuring the rate of rapid reactions by *in situ* molecular methods has limited applications so far. Recently, quick scanning XAS is used widely which is a synchrotron-based method to obtain and assess X-ray absorption image in near edge elemental structures and also fine elemental structures within a time frame of milliseconds. Q-XAS was first employed for the measurement of arsenic oxidation kinetics on hydrous magnesium oxide (HMO). This was materialized by a monochromator with cam operated continuous scanning facility by National Synchrotron Light Source (NSLS), USA, and this can collect all the edge and fine structures within 300 to 600 ms at even low concentrations and rendering highly sensitive images [34]. Time resolution of XAS measurements was prominently enriched by the oscillating movement of the monochromator with X-ray fluorescence spectra from high to low energy and vice versa. Q-XAS was also used to assess the metal adsorption and precipitation phenomenon and oxidation-reduction kinetics by utilizing flow and batch equipment during recent years.

Chromium is extensively present in two of its oxidation phases viz. Cr (III) and Cr (VI) in most usual conditions of the environment. Both the phases are characterized by opposite nature in terms of toxicity and finding the type of oxidation phase becomes so important in order to assess the soil performance. Many researchers used Q-XAS method to determine the initial rapid kinetics of Cr (III) phase oxidation upon HMO. This was carried out by measuring the spectrum of soil using XAS in 0.75 s and at pH values of 2 to 3.5 in the interval of 0.5 with Cr (III) concentrations between 40 and 100 mM. Cr (VI) phase of chromium results in soil as a result of continuous oxidation of Cr (III) oxidation on the HMO surface whose content could be measured by the edge display of this phase in XAS technique. Measurement of Cr (III) alone is sufficient as the other phase presence could be confirmed by measuring the former one. Figure 4.2 illustrates the kinetics of oxidation of Cr (III) that resulted in formation of Cr (VI) which could be observed in change in time intensity of pre-edge

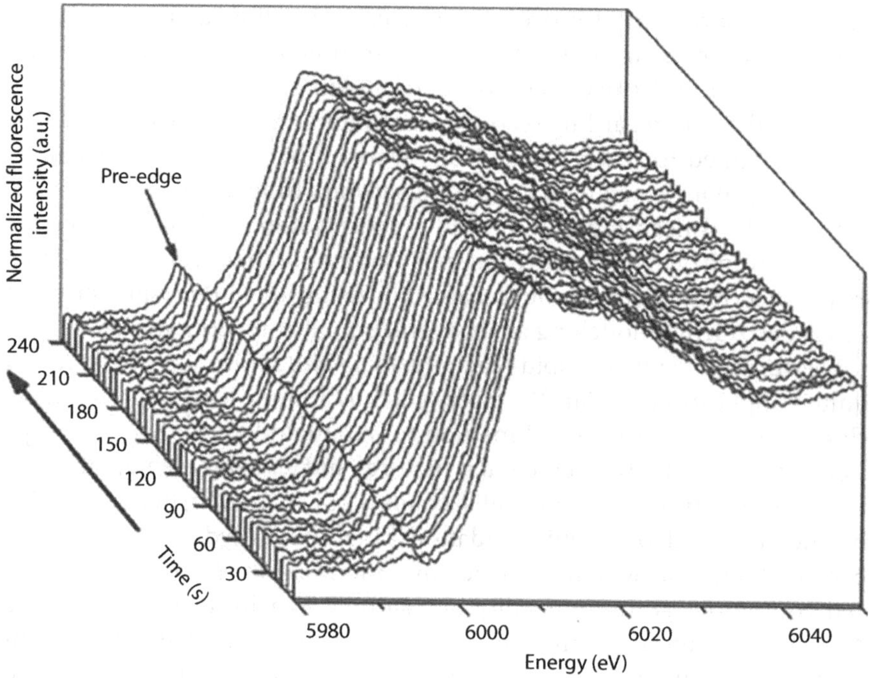

Figure 4.2 Q-XAS image of oxidation kinetics of a soil sample [32].

feature. It could also be observed from the figure that the oxidation of Cr (III) is very fast during the initial time of 120 s and at a concentration of 35 mM Cr (VI) was obtained [35]. Presence of Cr (VI) could be confirmed by measuring the pre-edge feature height of each peak reaction at a time range of 0–1 min. Some studies suggests that rate concentration was almost same for any given value of pH and at various concentration of HMO and Cr (III). Hence, advanced methods like Q-XAS provides the value of constants during initial reactions and also the values of chemical kinetics rate constants which, in turn, renders numerous perceptions on mechanism of reactions. But, other methods may not take the transportation process of molecules into account and so the obtained values are not rate constants but simple coefficients which have least significance in calculation of reaction parameters.

4.2.2 Carbon Speciation in Soils

Soil carbon which is present in the soil at almost equal to 50% of its contents finds a deeper connection with the biogeochemical process of soil.

These reactions may have greater influence over the bio-availability, fate, toxicity, organic-chemicals, transport, and metal and metalloids present in the soils. In spite of presence of excessive studies on the chemistry and reaction mechanisms of soils, its clear interaction with the other components remains a paradox for many decades. This inconsistency in understanding the behavior of soil chemistry is majorly due to the heterogeneity of substances, climatic change, and major content of carbon in soil [36]. However, apprehension of reactions is very important at least in terrestrial environment cycles. Such understandings could be enhanced by using many synchrotron-based methods like scanning transmission X-ray microscopy (STXM) and carbon near-edge X-ray absorption fine structure spectroscopy (C-NEXAFS) which could possible speciate the presence of carbon in soils so that the mechanisms and integration of current soil elements with other elements could be grasped easily.

Carbon impounding in soils can be easily determined by many reactions arising from iron hydroxides, clay minerals, and other metal hydroxides. Some good researches were carried out to specify the carbon sequestration on the soils were also briefed here. Researchers used C-NEXAFS and STXM methods to determine the carbon speciation in the soils of Christina river basin critical zone observatory (CRB-CZO) along with the analysis of association of clay fraction in those soils. Combining the above stated methods several carbon present regions were positioned along with the salient functional groups as shown in Figure 4.3. In the same study, correlation between Al and Fe elements was established stating that the oxides of iron were coated over the minerals of clay and structural iron elements. It was also stated that carbon and calcium present in the soil was strongly correlated which could be observed by the adsorption of positive charged

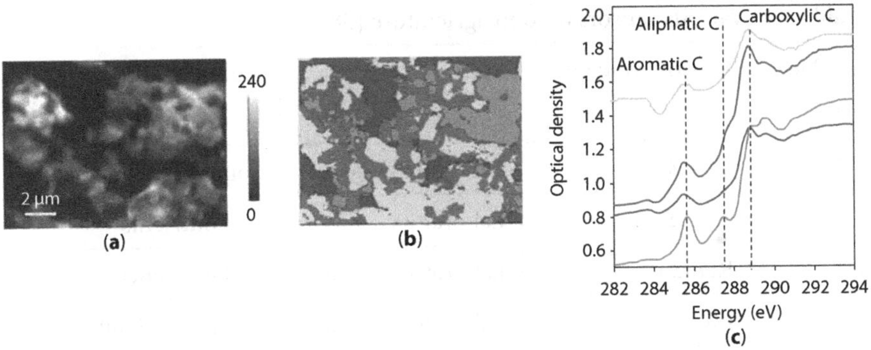

Figure 4.3 Micro and nanoscale characterization of species in the clay fraction of a soil [37].

calcium minerals into negative charge soil molecules. On contrary, poor correlation could be observed between carbon and silicon molecules since the silicon is comprised of quartz which has very low reactivity with carbon. From the study, it could be also concluded that no inorganic carbon was present in the sampled soil [37].

4.2.3 Soil Analysis Using Sensors

Many researchers stated the recent developments in sensor applications coupled with artificial intelligence (AI) for agricultural field by assessing the gap of knowledge. Initially, the precision agriculture field used ground buried sensors for the identification of organic matter in the soil, but nowadays, satellite, aircrafts, and vehicle-mounted sensors were employed. Table 4.1 shows various sensors and their applications in the field of agriculture.

4.2.4 Soil Analysis Using AI

AI techniques like aerial mapping can be easily used to create soil-moisture map instead of using sensors buried in land and the data from the aerial map can be integrated with soil moisture meter data for increasing the accuracy of those data. Various researchers predicted that the results of analysis by AI methods regarding weather forecasts and soil water balance would be helpful for irrigation and can be useful in taking precautionary measures against soil erosion or sudden floods. Forecasting the soil health in terms of carbon concentrations in soil may aid the farmers in knowing the health of the soil better and may execute their actions against the depletion of soil health avoiding the further effects. Internet of things (IoT) can

Table 4.1 Various sensors used in agriculture [38].

S. No.	Sensor concept	Status	Current results
1	Vis-NIR	Laboratory/Field	Soil pH and nutrients
2	Vis-MIR	Laboratory/Field	Soil mineral nitrogen
3	ATR Spec	Laboratory	Soil nutrients
4	Raman	Laboratory/Field	Soil nutrients
5	ISE	Laboratory/Field	Soil pH and nutrients
6	ISFET	Laboratory/Field	Soil pH and nutrients

also be used to forecast some advanced threats like plant pests, fungus, and insects like locusts [39].

4.2.5 Soil Analysis Using ML

Measurements of soil functions were converted into various soil properties by means of a function called pseudo-transfer function. Such properties are used to simulate and model soil properties in environmental science, hydrology, climate change impacts, soil research, exchange of carbon between soils, and the atmosphere and carbon cycle analysis to aid carbon farming. This function can also give input to many other research areas of soil including economic optimization, landscape design, and soil quality valuation. This function can also be used in integration with the machine learning (ML) functions for evaluating organic carbon content of the soil and electrical conductivity of the soil [40]. An integrated approach arises when the pseudo-transfer function and ML is coupled and the regional soil data were combined together thus eliminating the constraints of transferring the data from one region to the other. This integration is possible by general linear mixed model with inherent and available terms, predictor-frequency ranking system to forecast the errors in available models and residual maximum likelihood estimation. Soil information and measurement data along with monitoring data was obtained from a standard database and all the data models were tested from full model gradually by reducing the number of predictors. pH of the soil, geo-morphological mapping unit, particle size, and soil depth were the four major predictors used when organic carbon and electrical conductivity prediction methods were adopted. This coupled technique is said to have many advantages than any other conventional ML techniques. Figure 4.4 shows the generalized model of ML.

4.2.6 Soil Analysis Using IoT

Analysis of soil by methods like Arduino sensors and module, mobile application, web services, and Zigbee technology for designing the system may come under IoT techniques. In order to monitor the aquaculture water quality, a system of monitoring has been proposed to monitor properties like availability of oxygen, temperature, and pH. Such system model may also be used to enhance reduction of production cost and environmental control of soil. Arduino module senses temperature, pH, and oxygen availability in the soil with the help of sensors and the values were sent as database to the various cloud modules and web platforms. These data can be seen through

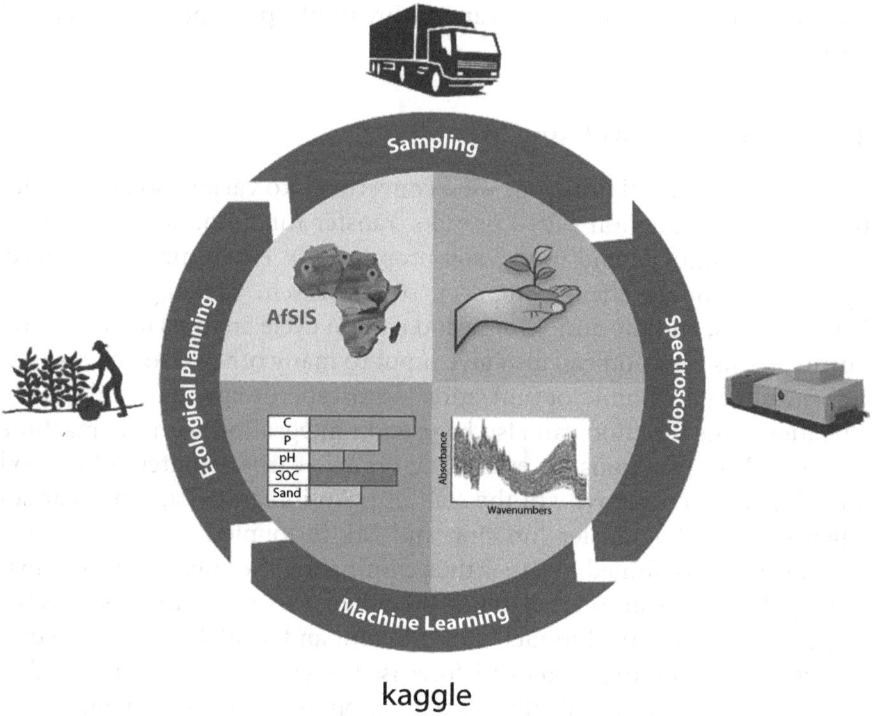

Figure 4.4 Soil analysis using machine learning [41].

mobile phones which are connected to the cloud modules and in near future alarm can be incorporated in the system to alert various functions with the aid of AI [42]. These systems were intended to enhance the farming productivity and vermin disease reduction. Some authors came up with the design of IoT-based agricultural system of production to forecast the production and growth of crops and also to saturate the supply and demand of agricultural products by evolving the necessary sensors. The majorly used various IoT-based sensors like soil electrical conductivity senor, temperature sensor, pH sensor, and humidity sensor. In recent time, advanced sensors like SEN: 0193, SEN: 0161, and DHT11are used to monitor soil conditions like moisture, pH, and temperature, respectively. This test is carried out by government authorized laboratories and the results are interpreted using t-test to find the level of significance between actual measured and laboratory values. This is very much useful for agriculture scientists, farmers, IoT experts, chemists, and agriculture professionals while this model could be further expanded based on geographical area to find electrical conductivity and other salient properties [43]. Figure 4.5 illustrates the uses of IoT in agriculture.

Figure 4.5 IoT in agriculture [43].

4.2.7 Soil Analysis Using Big Data

A national wide health identity card for soil was issued to each farmer by Government of India as a part of intimating them the status of the soil for the nutritious elements present in them and also about the nature of fertilizers farmers can use upon those soils for multiplying the yield. An initiative was taken by the government agricultural and research wing itself for analyzing the collected samples for the presence of salient 12 nutrients which were also printed on the soil health cards (SHCs) along with the fertilizers endorsed for the lands of farmers [44]. Nevertheless, this initiative has its own disadvantages listed as follows:

- Challenges was faced in operation of the testing facilities in the form of transportation of collected samples. Since the samples were collected at an interval of 2 hectare² irrigated lands and mixed together before transportation which were then analyzed in wet chemistry labs, the soil analysis standard remains a major question. Results generated by wet chemistry labs which were entered in SHC and the results of gold standard labs were found to have very low similarities which posed so many difficulties in operation.
- Nutrient recommendations and fertilizer choices were not accurately given by the current scheme of tests as yield of crops not only depend on nutrient's presence but also on many other factors including temperature of the soil, rainfall,

production practices, and soil acidity. Accurate yield response can be predicted only by taking all the above data into consideration which the current system does not possess.

- Finally, such big data has many other merits other than just being printed on cards as simple information. Such rigorously collected data can be released to start-up organizations, researchers, and to state government departments. They may use the data collected from millions of samples for correlating them with climate and irrigation information that may alert the farmer about major advises regarding the soil through mobile phone messages [45].

If these areas were clearly addressed by the current scheme, Indian agriculture may face an uphill in near future. A large database containing the soil information system measured from millions of samples may be created so that space related issuers for storing such large quantities of samples can be eradicated and the data can be retrieved, as per requirements, at a very swift rate than that of current time in cheap cost [46]. In spite of reducing the sowing depth, a farmer can be asked to utilize a tailored fertilizer recommended by the database. Fertilizer-based industries can use the big data to affirm the data regarding the soil and weather and may come up with tailored patterns of fertilizer blends required for various farmers from various districts. This may give an in-depth innovation in the field of agriculture. The process of analyzing data is illustrated in Figure 4.6.

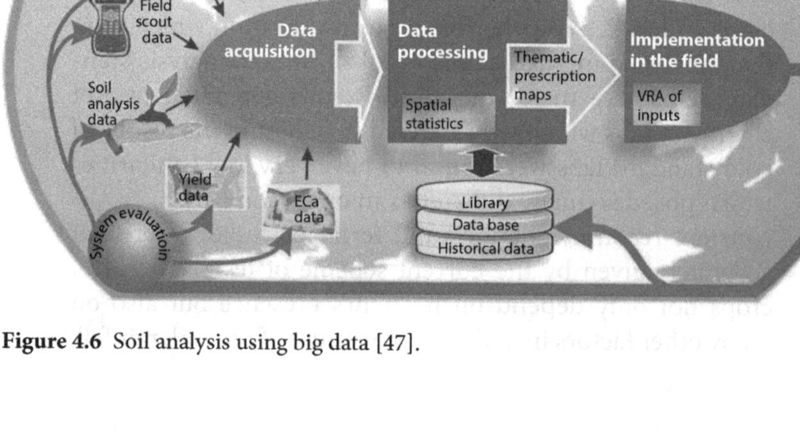

Figure 4.6 Soil analysis using big data [47].

4.2.8 Soil Analysis Using Drone and/or Satellite

Investigative and surveillance methods used for collecting crop and soil exploration data which will be used to train dynamic models. These analysis models may be used to unveil the effect of method used for enhancing the yield of crops and residual analysis development. These models cannot be developed by using the data acquired by satellites as they do not form the complete part of the data since they were measured by covering the entire earth surface in just 2 weeks. It is a highly time consuming process as compilation of those data is difficult due to their low resolution in the order of 5–80 ft per pixel [48]. There is also a possibility of missing data as they might be collected in cloudy weather conditions. On contrary, all the above disadvantages could be overcome by the use of drones for collecting the data. The prime advantages of using drones is that they are cheap method of collecting data, they are available on demand at any time required and the handling of collected images is also very simple. Drone recorded images possess a spatial resolution of 0.25–4.1 in per pixel which is much more helpful in assessing the soil condition and performance in small areas of field. Such method of assessment eases the decision of fertilizer combination and blend which adds much more values to the lands of farmers. Usage of hyper-spectral cameras has now become an inseparable part of using drones. Unlike traditional red, green, blue (RGB) cameras working in electro-magnetic spectrum, the hyper-spectral cameras work in infrared closer wavelengths for providing additional channels thus capturing even the rarest of images that were not identified by conventional RGB cameras. Such images may highly influence the performance assessments of crops since they provide information regarding pigmentation also. Figure 4.7 shows the schematic of communication method used for the data dissemination of soil.

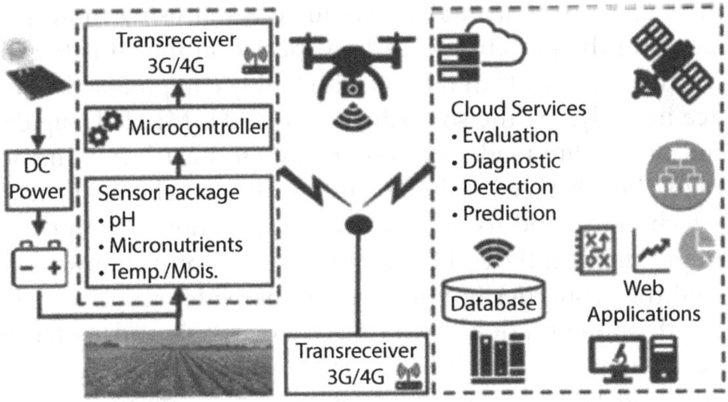

Figure 4.7 Communication model for data dissemination [49].

Plants emitting lower energy levels and lower leaf moisture content could be identified by capturing the amount of chlorophylls present on their leaves while healthy plants emits high energies and by shift of spectral content respectively which gives the information about the water shortage content in the plants. Some additional plant features like leaf shape and texture could be captured by drone images while this is not possible with low resolution imaging methods. Leaf shape is an important parameter to record since it influences not only the physiological and bio-chemical properties of the plant but also to identify and segregate the plant diseases without compromising the spectral content of the plant. Such data may render the quantum of nutrients present in the soil as solutions and exchange phases. Some of the specific measurements using drones may include Horiba LAQUA twin pocket meters and iMETOS MobiLab analyzers [50]. Such instrument may be used to record the soil data in monthly frequency which can be used to estimate the amount of nutrients produced within the soil itself after analyzing the gas losses. The methods used and the instruments for the above methods are of low cost and high accuracy. MobiLab can be used to measure K^+, NO^{3-}, and PO^{4-} and the pocket meters are capable of monitoring temperature, soil pH, conductivity, and moisture level of the soil. Using these measurement methods precision farming techniques can be developed for the farmers by combining physical soil characteristics and data regarding the soil fertility [51, 52].

4.3 Conclusion

This chapter dealt with the advances in soil chemistry which is based on the analysis of the soil at *in situ* level for the soil using molecular scale techniques, and the presence of carbon speciation in soil is the methods to find out the ingredients in the soil with handy equipment. The advances took place in analyzing the soil using sensors, AI, ML, IoT, big data, and drone and or satellite made the soil analysis to next level. The accuracy increases by the advanced technologies put the soil chemistry to develop the new techniques and to use for wider applications. The precise measurement put through the soil for continues monitoring allows the user to understand the properties in much better and wider scope. For the study of soil health and condition, use of big data analysis and AI has become wide spread along with the climatic change and carbon cycle. Use of non-traditional models which are the results of recent advancements in ML has also been frequent devoid of obvious models used generally. Most of the researchers are interested in using a simple and transparent ML algorithms which could easily render logical decisions and characteristics that were the results of traditional algorithms. This may lead to exploration of the soil resources in lot more verticals and utilizing them for maximizing the benefits of agriculture.

References

1. Bertsch, P.M. and Hunter, D.B., Applications of Synchrotron-Based X-ray Microprobes. *Chem. Rev.*, 101, 1809–1842, 2001.
2. Brown, G.E., Jr and Parks, G.A., Sorption of Trace Elements on Mineral Surfaces: Modern Perspectives from Spectroscopic Studies, and Comments on Sorption in the Marine Environment. *Int. Geol. Rev.*, 43, 963–1073, 2001.
3. Fenter, P.A., Rivers, M.L., Sturchio, N.C., Sutton, S.R. (Eds.), *Applications of Synchrotron Radiation in Low-Temperature Geochemistry and Environmental Sciences*, vol. 49, pp. 1–115, Geochemical Society, Mineralogical Society of America, Washington, D.C., USA, 2002.
4. Fenter, P.A., Rivers, M.L., Sturchio, N.C., Sutton, S.R. (Eds.), *Applications of Synchrotron Radiation in Low-Temperature Geochemistry and Environmental Sciences*, Geochemical Society, Mineralogical Society of America, Washington, D.C., USA, 2002.
5. Fenter, P.A., Rivers, M.L., Sturchio, N.C., Sutton, S.R. (Eds.), *Applications of Synchrotron Radiation in Low-Temperature Geochemistry and Environmental Sciences*, vol. 49, pp. 341–428, Geochemical Society, Mineralogical Society of America, Washington, D.C., USA, 2002.
6. Ulery, A.L. and Drees, L.R. (Eds.), *Methods of Soil Analysis Part 5, Mineralogical Methods*, pp. 387–463, Soil Science Society of America, Madison, WI, USA, 2008.
7. Lombi, E. and Susini, J., Synchrotron-based techniques for plant and soil science: Opportunities, challenges and future perspectives. *Plant Soil*, 320, 1–35, 2009.
8. Singh, B. and Gräfe, M. (Eds.), Synchrotron-Based Techniques in Soils and Sediments, in: *Developments in Soil Science*, vol. 199, p. 254, Elsevier, New York, NY, USA, 2010.
9. Ginder-Vogel, M. and Sparks, D.L., Synchrotron-based Techniques in Soils and Sediments, in: *Developments in Soil Science*, B. Singh and M. Gräfe (Eds.), pp. 1–26, Elsevier, New York, NY, USA, 2010.
10. Gräfe, M., Donner, E., Collins, R.N., Lombi, E., Speciation of metal(loid)s in environmental samples by X-ray absorption spectroscopy: A critical review. *Anal. Chim. Acta*, 822, 1–22, 2014.
11. Sparks, D.L. (Ed.), *Advances in Agronomy*, vol. 126, pp. 1–148, Academic Press, San Diego, CA, USA, 2014.

12. Zhang, P.C. and Sparks, D.L., Kinetics and Mechanisms of Sulfate Adsorption/Desorption on Goethite Using Pressure-Jump Relaxation. *Soil Sci. Soc. Am. J.*, 54, 1266–127.
13. Hayes, K.F., Roe, A.L., Brown, Jr., G.E., Hodgson, K.O., Leckie, J.O., Parks, G.A., *In Situ* X-ray Absorption Study of Surface Complexes: Selenium Oxyanions on alpha-FeOOH. *Science*, 238, 783–786, http://www.jstor.org/stable/1700357, 1987.
14. Siebecker, M.G., Chaney, R.L., Sparks, D.L., Nickel speciation in several serpentine (ultramafic) topsoils via bulk synchrotron-based techniques. *Geoderma*, 298, 35–45, 2017.
15. Scheidegger, A.M. and Sparks, D.L., Kinetics of the formation and the dissolution of nickel surface precipitates on pyrophyllite. *Chem. Geol.*, 132, 157-164, 1996.
16. Violante, A., Huang, P.M., Gadd, G.M. (Eds.), *Biophysico-Chemical Processes of Heavy Metals and Metalloids in Soil Environments*, pp. 97–124, Wiley, New York, NY, USA, 2008.
17. Chen, C., Dynes, J., Wang, J. *et al.*, Soft X-ray Spectromicroscopy Study of Mineral-Organic Matter Associations in Pasture Soil Clay Fractions. *Environ. Sci. Technol.*, 48, 12, 6678–6686, 2014.
18. Vasisht, D., Kapetanovic, Z., Won, J., Jin, X., Chandra, R., Sinha, S.N., Kapoor, A., Sudarshan, M., Stratman, S., FarmBeats: an IoT platform for datadriven agriculture, in: *Proceedings of the 14th USENIX Symposium on Networked Systems (NSDI'17)*, Boston, MA, 27–29 March 2017, pp. 515-529, 2017.
19. Morgan, R.P.C. and Nearing, M., *Handbook of erosion modelling*, John Wiley & Sons, Chichester, 2016.
20. Belayneh, A. and Adamowski, J., Drought forecasting: artificial intelligence methods, in: *Exploring natural hazards*, Bartlett, D. and Singh, R. (Eds.), pp. 207–224, 2018.
21. Ehsani, M.R., Upadhyaya, S.K., Slaughter, D., Shafii, S., Pelletier, M., A nir technique for rapid determination of soil mineral nitrogen. *Precis. Agric.*, 1, 219, 1999.
22. Mirkouei, A., A Cyber-Physical Analyzer System for Precision Agriculture. *J. Environ. Sci. Curr. Res.*, 3, 016, 2020.
23. Ehsani, M.R. *et al.*, "Quantitative measurement of soil nitrate content using midinfrared diffuse reflectance spectroscopy." Quantitative Measurement of Soil Nitrate Content Using Mid-infrared Diffuse Reflectance Spectroscopy, American Society of Agricultural Engineers (ASAE), Milwaukee, Wisconsin, p. 1, 2000.
24. ViscarraRossel, R.A., Walvoort, D.J.J., McBratney, A.B., Janik, L.J., Skjemstad, J.O., Visible, near infrared, mid infrared or combined diffuse reflectance spectroscopy for simultaneous assessment of various soil properties. *Geoderma*, 131, 59, 2006.

25. Shibusawa, S., On-line real time soil sensor. *IEEE/ASME International Conference on Advanced Intelligent Mechatronics (AIM 2003)*, vol. 2, IEEE, p. 1061, 2003.
26. Christy, C.D., Drummond, P., Laird, D.A., An on-the-go spectral reflectance sensor for soil. *ASAE Annual Meeting, 2003*, American Society of Agricultural and Biological Engineers, p. 1, 2003.
27. Jahn, B.R., Linker, R., Upadhyaya, S.K., Shaviv, A., Slaughter, D.C., Shmulevich, I., Mid-infrared spectroscopic determination of soil nitrate content. *Biosyst. Eng.*, 94, 505, 2006.
28. van Staden, J.K.F., Nuta, R.-G., Arnold, G.-L.T., Determination of nitrite from water catchment areas using graphite based electrodes. *J. Electrochem. Soc.*, 165, B565, 2018.
29. Tyszczuk-Rotko, K. and Jedruchniewicz, K., Ultra sensitivesensor for uranium monitoring in water ecosystems. *J. Electrochem. Soc.*, 166, B837, 2019.
30. Lobsey, C. *et al.*, An automated system for rapid in-field soil nutrient testing. *Soil Solutions for a Changing World, 19th World Congress of Soil Science Brisbane*, Australia, August 2010, pp. 1–6, 2010.
31. Lobsey, C.R., Viscarra Rossel, R.A., Mcbratney, A.B., Proximal soil nutrient sensing using electrochemical sensors, in: *Proximal Soil Sensing*, p. 77, Springer, Dordrecht, 2010.
32. Kim, H.J., *Ion-selective electrodes for simultaneous real-time analysis for soil macronutrients*. Ph.D. Thesis, University of Missouri-Columbia, Columbia, United States, 2006.
33. Aravamudhan, S. and Bhansali, S., Development of micro-fluidic nitrateselective sensor based on doped-polypyrrole nanowires. *Sens. Actuators B*, 132, 623, 2008.
34. Soriano-Disla, J.M., Janik, L.J., ViscarraRossel, R.A., Macdonald, L.M., McLaughlin, M.J., The performance of visible, near-, and mid-infrared reflectance spectroscopy for prediction of soil physical, chemical, and biological properties. *Appl. Spectrosc. Rev.*, 49, 139, 2014.
35. Sudduth, K.A., Hummel, J.W., Birrell, S.J., Sensors for site-specific management, in: *The State of Site-specific Management for Agriculture*, vol. 10, p. 183, ASA, CSSA, SSSA, Madison, WI, 1997.
36. Adamchuk, V., II, Lund, E.D., Sethuramasamyraja, B., Morgan, M.T., Dobermann, A., Marx, D.B., Direct measurement of soil chemical properties on-the-go using ion-selective electrodes. *Comput. Electron. Agric.*, 48, 272, 2005.
37. Fotovatikhah, F., Herrera, M., Shamshirband, S., Chau, K.W., FaizollahzadehArdabili, S., Piran, M.J., Survey of computational intelligence as basis to big flood management: Challenges, research directions and future work. *Eng. Appl. Comput. Fluid Mech.*, 12, 411–437, 2018.
38. Yan, Y., Feng, C.C., Wan, M.P.H., Chang, K.T.T., Multiple regression and artificial neural network for the prediction of crop pest risks, in: *International*

conference on information systems for crisis response and management in Mediterranean countries, pp. 73–84, 2015.

39. Benke, K.K., Norng, S., Robinson, N.J., Chia, K., Rees, D.B., Hopley, J., Development of pedotransfer functions by machine learning for prediction of soil electrical conductivity and organic carbon content. *Geoderma*, 366, 114210, 2020 May 1.
40. Encinas, C., Ruiz, E., Cortez, J., Espinoza, A., Design and implementation of a distributed IoT system for the monitoring of water quality in aquaculture, in: *2017 Wireless Telecommunications Symposium (WTS)*, IEEE, 2017.
41. Coyle, S., Lau, K.-T., Moyna, N., O'Gorman, D., Diamond, D., Di Francesco, F. *et al.*, BIOTEX—Biosensing textiles for personalised healthcare management. *IEEE Trans. Inf. Technol. Biomed.*, 14, 2, 364–370, 2010.
42. Yamaguchi, N., Sakai, Y., Shiraishi, T., Onishi, S., Kowata, T., E-kakashi project, an agri sensor network using ad hoc network technology, in: *2011 Proceedings of SICE Annual Conference (SICE)*, IEEE.
43. Lee, M., Hwang, J., Yoe, H., Agricultural production system based on IoT, in: *2013 IEEE 16th International Conference on Computational Science and Engineering (CSE)*, IEEE, 2013.
44. Kaur, P. and Bharti, V., Applications of Smart Devices, in: *Recent Trends and Advances in Artificial Intelligence and Internet of Things 2020*, pp. 483–517, Springer, Cham, 2020.
45. Jiang, Y., Li, C., Robertson, J.S., Sun, S., Xu, R. *et al.*, GPheno Vision: A ground mobile system with multi-modal imaging for field-based high throughput phenotyping of cotton. *Sci. Rep.*, 8, 1213, 2018.
46. Wahabzada, M., Mahlein, A.-K., Bauckhage, C., Steiner, U., Oerke, E.-C. *et al.*, Plant phenotyping using probabilistic topic models: Uncovering the hyperspectral language of plants. *Sci. Rep.*, 6, 22482, 2016.
47. Van Evert, Frits, K., Fountas, S., Jakovetic, D., Crnojevic, V., Travlos, I., Kempenaar, C., Big data for weed control and crop protection. *Weed Res.*, 57, 4, 218–233, 2017.
48. Lottes, P., Khanna, R., Pfeifer, J., Siegwart, R., Stachniss, C., UAV-based crop and weed classification for smart farming. *IEEE International Conference on Robotics and Automation (ICRA)*, IEEE, Singapore, 2017.
49. Khan, M.J., Khan, H.S., Yousaf, A., Khurshid, K., Abbas, A., Modern trends in hyperspectral image analysis: A review. *IEEE Access*, 6, 14118–14129, 2018.
50. King, A., The future of agriculture. *Nature*, 544, 21–23, 2017.
51. Shaw, G.A. and Burke, H.K., Spectral imaging for remote sensing. *Linc. Lab. J.*, 14, 3–28, 2003.
52. Mulla, D.J., Twenty five years of remote sensing in precision agriculture: Key advances and remaining knowledge gaps. *Biosyst. Eng.*, 114, 4, 358–371, 2013.

5

An Overview of Soil Chemistry and Role of Its Components in Sorption of Heavy Metals in Soil

Shagufta Jabin[1]*, Jamal A. Khan[2] and Sapana Jadoun[3]

[1]*Department of Chemistry, Faculty of Engineering & Technology, Manav Rachna International Institute of Research and Studies, Faridabad, Haryana, India*
[2]*Department of Applied Chemistry, Faculty of Engineering & Technology, Aligarh Muslim University, Aligarh, U.P, India*
[3]*Department of Chemistry, Lingaya's Vidyapeeth, Faridabad, Haryana, India*

Abstract

All basic needs of human beings and other living organisms are fulfilled by soil. It is one of the most important resources of universe. Interest in soil chemistry developed due to its unique property to produce and sustain crops. The physical and chemical properties of soil are very important to understand because soil productivity depends upon these factors. The important parameters of physico-chemical properties of soil are pH, temperature, electrical conductivity, and cation exchange capacity. Composition of soil is also one of the driving forces to understand the chemistry of soil. Solid, liquid, and gaseous phase of soil gives complete idea about the quality of soil. Solid phase contain inorganic and organic components. Kinetic reactions of soil mostly depend upon inorganic components. Soil organic matter has potential to influence soil chemistry. Soil characteristics mainly include soil structure, soil color, bulk density, and particle size distribution. They are key indicators to understand the chemistry of soil. Soil has unique property to be used as adsorbent for removal of heavy metals from aqueous medium. The use of clay and other inorganic minerals of soil in natural and modified form for removal of heavy metals has been discussed extensively in this chapter. Keeping in view the applicability of soil for removal of heavy metals, it is expected that this technique can be applied for further treatment in variety of water and wastewater.

**Corresponding author*: shaguftaajabin@gmail.com

Inamuddin, Mohd Imran Ahamed, Rajender Boddula and Tariq Altalhi (eds.) Applied Soil Chemistry, (79–104) © 2021 Scrivener Publishing LLC

Keywords: Kinetic reactions, soil organic matter, cation exchange capacity, temperature, electrical conductivity, heavy metals

5.1 Introduction

The well-being of mankind mainly depends upon environment and ecological factors. However, ecological factors and environment depend upon the quality of soil available on the planet [1]. The term soil has been originated from "solum" a Latin word. Curiosity about soil developed because of its properties of sustaining crops and capacity to produce. Soil plays a significant role in sustaining life in universe [2, 3]. Soil provides medium for adsorption of pollutants, microorganisms, and excess water [4]. Soil is directly derived from degradation and weathering of rocks. From the standpoint of soil science, soil is having complex chemistry, containing different composition and definite structure. It consists of organic and inorganic composition having definite chemical, physical, and biological properties.

5.2 Composition of Soil

Composition of soil is divided into different categories. They are mainly present in solid, liquid, and gaseous form in different proportion [5]. The composition of an ideal soil has been shown in Figure 5.1.

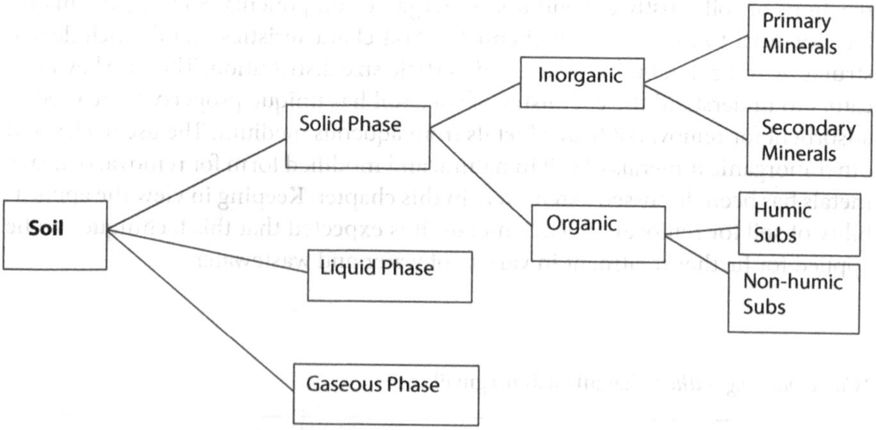

Figure 5.1 Composition of soil.

5.2.1 Solid Phase

Composition of solid phase of soil consists of following matters.

5.2.1.1 Soil Inorganic Matters

The inorganic matters in soils are made up of minerals of various kinds and small rock fragments which have been formed by physical and chemical weathering. The inorganic components of soil are mostly present in solid form. Inorganic components mostly influence kinetic reactions and other processes in soil. Inorganic materials comprise both primary and secondary minerals. These minerals vary in size. Primary minerals have not been changed chemically since its deposition. The different kind of inorganic matters of soil along with its chemical formula have been tabulated in Table 5.1 [4].

Table 5.1 Different components of soil and its chemical formula.

Name	Chemical formula
Quartz	SiO_2
Muscovite	$KAl_2(AlSi_3O_{10})(OH)_2$
Biotite	$K(Mg, Fe)_3(AlSi_3O_{10})(OH)_2$
Feldspars Orthoclase Microcline Albite	 $KAlSi_3O_8$ $KAlSi_3O_8$ $NaAlSi_3O_8$
Amphiboles Tremolite	 $Ca_2Mg_5Si_8O_{22}(OH)_2$
Pyroxenes Enstatite Diopside Rhodonite	 $MgSiO_3$ $CaMg(Si_2O_6)$ $MnSiO_3$
Olivine	$(Mg, Fe)_2SiO_4$
Epidote	$Ca_2(Al, Fe)_3Si_3O_{12}(OH)$
Tourmaline	$(Na, Ca)(Al, Fe^{3+}, Li, Mg)_3Al_6(BO_3)_3(Si_6O_{18})(OH)_4$
Zircon	$ZrSiO_4$
Rutile	TiO_2

The secondary minerals are mostly present in clay form and it is smallest material of soil. Clay is the most active portion of soil. The smallest clay particle is having colloidal properties. The colloidal nature of soil impacts properties of soil significantly. The size of clay particles is generally less than 0.002 mm. According to soil chemistry, they have been classified into five different categories:

- Kaolinite
- Montmorillonite
- Illite
- Chlorite
- Amorphous Clays

5.2.1.1.1 Kaolinite

The term "Kaolinite" was first proposed by "Johnson and Blake" in 1867. The chemical formula of kaolinite is $Si_4Al_4O_{10}(OH)_8$. The structure of Kaolinite is based on single tetrahedral sheet topped by a slightly distorted gibbsite sheet, both being joined by oxygen bonding through condensation and splitting of water betw een adjoining hydroxyl groups in vertical position [6]. Kaolinite received great attention because it can adsorb a variety of pollutants [7]. In conventional chemical treatment, kaolinite is utilized as major adsorbents in waste water treatment [8]. Kaolinite is having low swelling capacity and low moisture content. The size of the unit cell of kaolinite is 7.2 A°. The cation exchange capacity (CEC) of the kaolinite is very low and present in the range of 3 to 15 meq/100 g of clay. The structure of kaolinite is shown in Figure 5.2.

5.2.1.1.2 Montmorillonite

The montmorillonite was first proposed by Le Chatellier. The chemical formula of montmorillonite is $Mx\,(Al, Fe^{2+}, Mg)_4Si_8O_{20}(OH)_4$ where M is interlayer metal cation. It is composed of octahedral sheet of alumina bounded between two silica sheets. These layers are joined together by strong primary valence forces due to sharing of oxygen atoms in vertex position [9, 10]. It holds high CEC which varies in the range of 80 to 120 meq/100 g of clay. Montmorillonite is used as potential adsorbent for removal of heavy metals and other pollutants from aqueous medium [11, 12]. The structure of montmorillonite is shown in Figure 5.3.

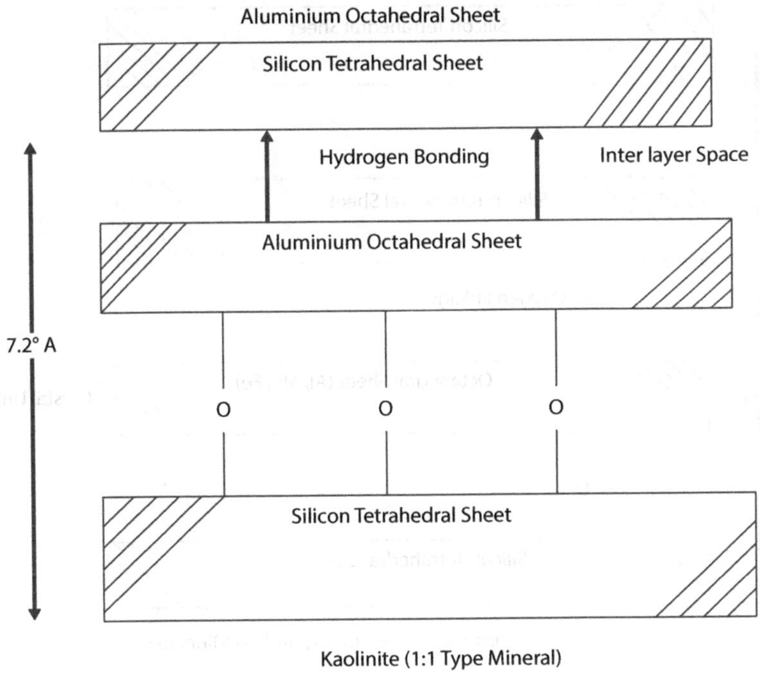

Figure 5.2 Structure of kaolinite.

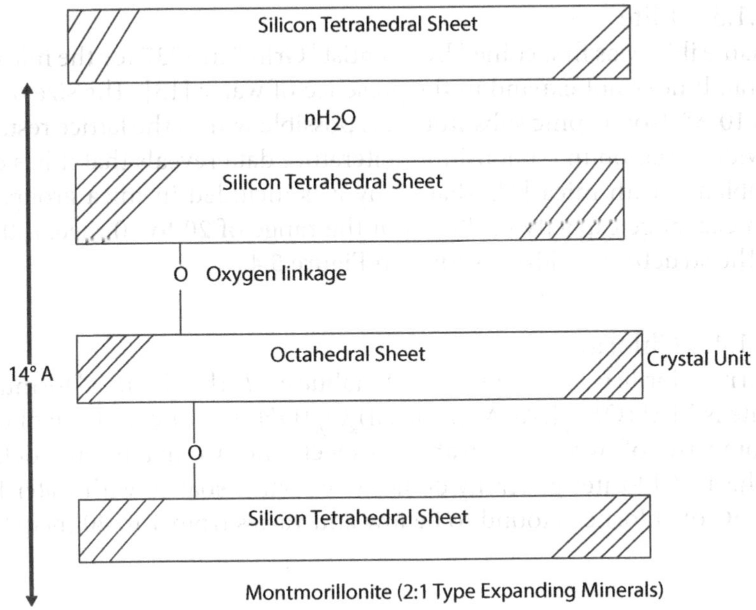

Figure 5.3 Structure of montmorillonite.

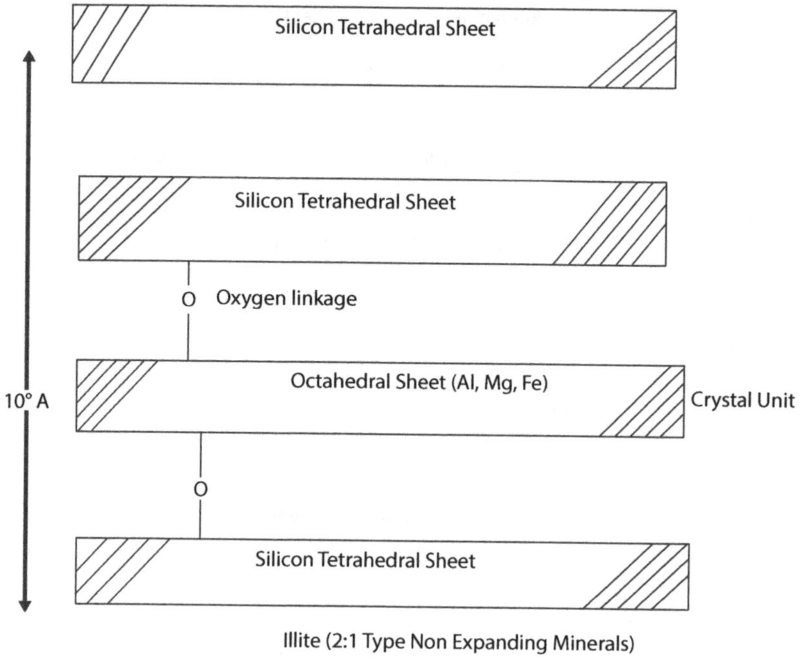

Figure 5.4 Structure of illite.

5.2.1.1.3 Illite

The name illite was first coined by scientist "Grim" in 1937 for the mica like mineral. It does not expand in the presence of water [13]. The size of unit cell is 10 A°. Isomorphic substitution is possible within the lattice resulting in a wide variation in composition. Literature data reveals that it has close resemblance with mica [1]. That's why it is included in mica group. The cation exchange capacity of illite is in the range of 20 to 40 meq/100 g of clay. The structure of illite is shown in Figure 5.4.

5.2.1.1.4 Chlorite

The term chlorite was first given by Donahue *et al.* The chemical formula of chlorite is $MAl\ (OH)_6](Al, Mg)_4(Si, Al)_8O_{20}(OH, F)_4$. The unit cell of chlorite comprises of two silica tetrahedral sheets and two magnesium octahedral sheets. Chlorites generally do not swell when soaked with water [13]. The CEC of chlorite is found to be low and varies from 1 to 40 meq/100 g of clay.

Table 5.2 Chemical composition of soil inorganic matter [14].

Inorganic matter	SiO_2 (%)	Al_2O_3 (%)	Fe_2O_3 (%)	TiO_2 (%)	Cao (%)	NaO (%)	MgO (%)	K_2O (%)
Kaolinite	53.70	43.60	2.00	0.10	-	-	-	0.50
Montmorillonite	65.34	12.89	2.38	0.52	0.24	0.53	0.95	1.54
Illite	62.72	14.34	5.58	0.82	7.27	1.01	1.82	3.59
Chlorite	27.40	18.90	2.40	-	-	-	34.00	-
Amorphous clay (Indian)	48.12	34.54	2.48	0.40	0.83	-	0.50	-

5.2.1.1.5 Amorphous Clays

The constituents of amorphous clay have short range structure. Their chemical composition is variable and consists of alumina, silica, and hydrous oxides of iron and aluminum [14]. The properties of amorphous clay depend upon pH of soil. The close association of amorphous clay with soil organic matter (SOM) makes their separation and characterization a difficult task. Even small amounts of amorphous clay may have strong influence on chemical and physical properties of soil because of large and active surface area [15].

The chemical composition of different soil inorganic matter has been tabulated in Table 5.2.

5.2.1.2 Soil Organic Matter

It is one of the most active portions of the soil and it plays very important role in influencing soil chemistry [16]. It includes all the materials of plant, animal, and microbial content present in the soil. It is generally dark in color and is not having any unique structure but it has significant impacts in every soil reactions and processes. The composition of SOM is 3% to 6% in agriculture soil. It improves aeration, water infiltration, and water holding capacity of soil [17]. Stickiness of soil also reduces due to presence of organic matter. It is major source of Sulphur and phosphorus. It also contain significant amount of carbon (58%) along with small quantity of hydrogen, oxygen, boron, and molybdenum [18].

Table 5.3 Chemical composition of soil organic matter.

Organism	Ash (%)	Protein (%)	Hemicellulose protein (%)	Cellulose (%)	Lignin (%)	Lipid tannins (%)
Algae	20–30	10–15	50–60	5–10	Nil	1-3
Bacteria	2–10	40–70	Nil	Nil	Nil	1–40
Moss	3–10	5–10	30–60	15-25	Nil	5–10
Lichen	2–6	3–5	60–80	5-10	8–10	1–3
Fern	6–7	4–5	20–30	20-30	20–30	2–10
Deciduous						
Wood	0.1–1	0.5–1	20–30	40-50	20–25	5–15
Leaves	2–5	4–10	10–20	15-25	20–30	5–15
Conifers						
Stem	0.1–1	0.5–1	20–30	40–50	20–25	5–15
Needles	3–8	4–10	10–20	15–25	25–30	5–15

All carbon atoms present in it are bonded together by strong bond which is backbone of SOM [19]. SOM is having high specific surface and its CEC ranges from 150 to 300 Cmol/kg.

CEC of soil surface i s majorly due to organic matter. Because of CEC, organic matter is key sorbent of heavy metals, pesticides, plant macronutrients, and micronutrients. The details of sorption capacity of soil will be discussed in another section. SOM enhance the ability of a soil to resist pH change. That's why it is also called as buffering agent. Orlov *et al.* in 2005 studied chemical composition of SOM which has been tabulated in Table 5.3 [20].

SOMs are mostly present in the form of carbohydrates (cellulose, pectin, and hemicellulose), proteins, and lignin [21]. All living organisms convert into biomolecules after death and subsequently they convert into humic substances. The functional groups present in humic acid are phenolic, alcoholic, carboxylic, carbonyls, ring nitrogen, and secondary amino groups [22].

5.2.2 Liquid Phase

It is significantly present in the form of soil water. It converts plant nutrients into its ionic form.

Soil water and its salt concentration also control osmotic pressure of roots [23]. Water has various chemical and physical properties which results into molecular structure [24]. High dielectric constant and ionic dissociation of soil water are results of polar nature of molecules present in soil. Strong intermolecular force of attraction of water molecules results into high surface tension, heat of vaporization, heat capacity, and boiling point. Manahan in 2004 and Pierzynski in 2005 studied physical and chemical properties of liquid and their significance in soil chemistry [25, 26]. Their studies have been summarized and tabulated in Table 5.4.

Dielectric constant of soil water is significantly higher than solid and gaseous phase of soil. This unique property makes it a strong solvent [27].

Amphoteric nature of soil water is an important chemical property. Because of polar nature of soil solution, it readily dissociate into cation and anion.

$$H_2O_{(l)} \leftrightarrow H^+_{(aq)} + OH^-_{(aq)} \tag{5.1}$$

When it undergoes reaction with base, it acts as an acid and release proton.

$$H_2O_{(l)} + NH_3 \leftrightarrow NH^+_{4\,(aq)} + OH^-_{(aq)} \tag{5.2}$$

Table 5.4 Physical and chemical properties of soil water.

Physical & chemical properties	Characteristics/justification	Significance
High dielectric constant	Dipole moment with negative and positive charge helps to stabilize solutes.	Works as excellent solvent for charge species.
High heat capacity	Strong interactions exist between water molecules to enhance temperature	Small change in thermal energy does not change temperature.
High surface tension	Cohesion at interface leads to high surface tension.	Formation of capillary behavior
Ionic dissociation	Soil water ionizes into cation and anion due to the polar nature of molecules	Soil water exhibit acid base chemistry due to ionic dissociation.

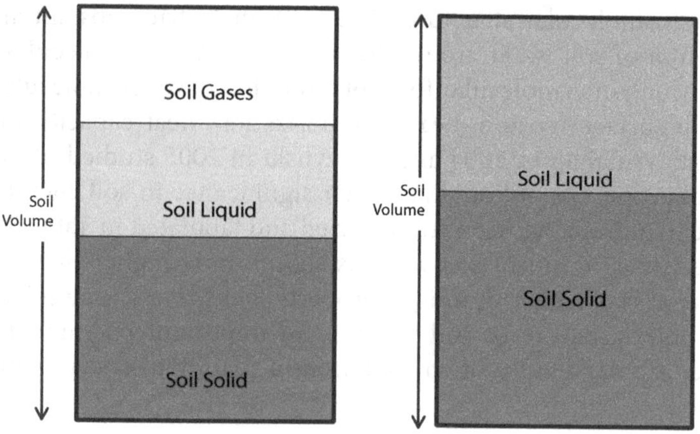

Figure 5.5 Saturated and unsaturated soil conditions [28].

However, when it reacts with acid, soil water acts as a base and accepts protons

$$H_2O_{(l)} + HCl \leftrightarrow H_3O^+_{(aq)} + Cl^-_{(aq)} \quad (5.3)$$

These reactions show that soil water is having amphoteric property which enables acid base property of soil. That's why it shows large degree of pH range in soil solution which facilitates plant growth. In saturated condition, soil liquid occupies approximate 50% soil volume. However, in unsaturated condition, liquid occupies 10%–20% soil volume as shown in Figure 5.5 [28].

5.2.3 Gaseous Phase

Gaseous phase of soil plays important role in influencing soil atmosphere and crop production. Soil gaseous phase primarily consist of oxygen, carbon di oxide, nitrogen, and water vapor. Oxygen is found to be most important gas because it is required by soil organisms. Gases are generally present between the air space of soil. The concentration of different gases in soil fluctuates from place to place [28]. It is important component of soil for survival of soil fauna and soil flora. Radon and methane are also included in soil gases [29]. They adversely affect environment and produce volatile organic compounds by mixing with petroleum hydrocarbon [30].

Gases are generally present in that space of soil which is not occupied by soil water. So, the composition of soil air is measured by the water content of the soil.

Percentage composition of atmospheric gases in soil is as follows:

(i) Nitrogen = 79.0%
(ii) O_2 = 20.97%
(iii) CO_2 = 0.03%

The inert gases are also included in nitrogen content. The soil air varies with the climatic condition and nature of soil [29]. It is evident from literature that carbondioxide in soil increases after rainfall [31]. It may be because of decomposition of organic matter. The percentage of carbon dioxide also increases in moist soil which leads growth of rice [31]. However, oxygen is important for all biological activities occurring in soil.

5.3 Soil Characterization

Soil characterization is generally analyzed by their structure, color, texture, consistence, and the amount of rocks, roots, and free carbonates they contain. Soil structure, bulk density, particle size distribution, pH, and soil fertility are important characteristics of soil [32]. Some of important characteristics of soil have been discussed in detail.

5.3.1 Structure

It refers to arrangement and organization of different particles of soil. Soil structure controls chemical, physical, and biological process which occurs in soil including movement and storage of soil water, gases, and heat [33]. It is one of the most complex soil properties because of heterogeneity in size and shape of soil particles. Due to variability from one point to another, a definite soil structure could not be explained. Interest in structure of soil was developed because of significance of nutrient storage and cycling. According to recent research, soil structure is also called as a store house of carbon and a large reservoir of atmospheric carbon dioxide [34]. On one hand, soil is having single grained structure with unsystematic arrangement of silt and clay. On the other hand, clay is having large massive structure.

Soil structure is always present in dynamic form which undergoes continuous change. Continuous change in the structure of soil at different levels makes it difficult to understand. Literature reveals that physical properties of soil are correlated with particle size distribution [35]. However, soil texture does not change appreciably with time [36]. From mechanical and chemical point of view, aggregation of soil is often confused with soil flocculation. However, soil aggregation is a kind of cementation of all particles without dispersion of particles in soil water [37]. Whereas, flocculation is an important physico-chemical operation for removal of suspension in the presence of flocculating agents. The mechanism of aggregation of soil is the subject of continuous review [38, 39]. A schematic diagram of soil aggregation is presented in Figure 5.6.

The different particles of soil combine together and lead to the formation of micro-aggregates. These micro-aggregates results into macro-aggregates. The size of macro-aggregates in a soil is dependent upon the kind of interaction between micro-aggregates. As per literature, micro-aggregates are stronger and denser than macro-aggregates [40]. It is due to smaller pores of micro-aggregates. In addition, outer layer of macro-aggregates is more porous and conductive as compared to inner layer.

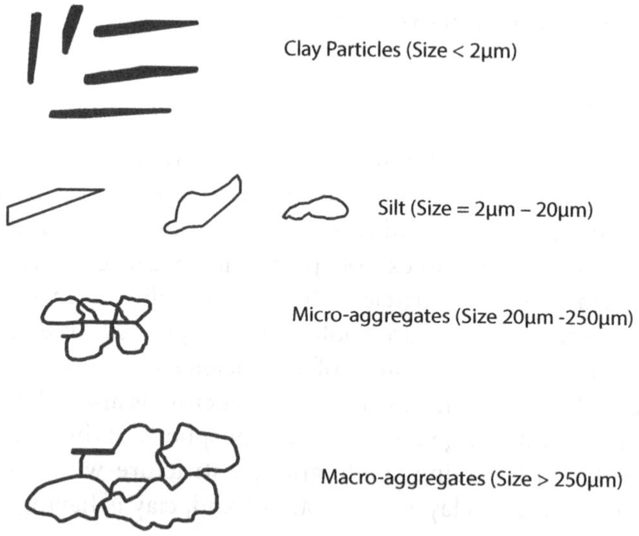

Figure 5.6 Conceptual model of soil structure [41].

Table 5.5 Association of soil color with elements.

Elements	Soil color
Cobalt	Blue/Red
Copper	Green/Blue
Zinc	Pale blue
Iron	Blue or Red depending upon ionic form
Lead	Yellow
Calcium	White
Chromium	Blue-Green
Vanadium	Green/Blue/Purple

5.3.2 Color

Soil color is mostly dependent on presence of elements, moisture content, and soil organic matter. Redish color in soil is generally associated with high iron content in soil; however, dark brown color is associated with large quantities of humus [41]. Further, soil age can also be predicted by soil color. Due to presence of calcium carbonate, soil appears white in color. It can be seen from the literature that strong relationship exists between physico-chemical characteristics of soil and color [42, 43]. Associations of soil color with different elements have been tabulated in Table 5.5 [44].

There are inter-relations involving between soil color and presence of elements [45]. But, concentration of any element is not solely responsible for color of soil. Further, clay minerals and SOM also have strong influence on soil color. The organic rich soil is having high fertility due to presence of organic content. Further, relationship between color and soil parameters has been studied by soil scientists especially for iron content and organic matter presence in soil [46].

5.3.3 Texture

Sand, silt, and clay are three main components of soil for determination of texture. Among all three, sand is largest in size followed by silt and clay. Clay is sticky part of soil and it helps in binding all particles of soil. The size of sand, silt, and clay in a soil sample is called the particle size distribution. However, clay particles undergo shrinking and swelling because of irrigation, evaporation, precipitation, and plant water uptake. So, the shrinking and swelling of clay particles also influence texture of soil [36].

5.3.4 Bulk Density

It is a measure of packing or compactness of the soil. In addition to the structure and composition of soil, bulk density also depends upon sand, silt, and clay particles and the volume of pore space. Compactness of soil is meant to bring all soil particles together. It results in decreasing the bulk volume and increasing the bulk density [47]. However, bulk density of soil varies in between 0.5 and 2.0 g/cm^3. Soil bulk density also helps to determine spatial arrangement which includes soil matrix and provide information about porosity, penetration resistance, and compactness of soil. In addition to it, it also provides information regarding rainfall-runoff, gas and heat exchange, root growth, and crop yield. Bulk density is not natural soil property but it depends very much on rainfall, wetting, and drying of soil and other natural and anthropogenic processes [48].

5.3.5 Particle Size Distribution

Particle size and its distribution are of great significance for shaping the properties of particulate matters. It is one of the most important physical properties of soil. Particle size analysis is normally done for the determination of mole fraction of sand, silt, and clay [49]. Exact texture of soil is determined by particle size distribution. There are wide ranges of techniques available for measurement of particle size distribution. Particle size distribution can directly be measured by microscope [50]. Few indirect methods are also available such as diffraction method [50, 51]. It is a mathematical function which explain amount of particles present according to their size. Separation of soil particles by sieve analysis is also a process adopted to assess the particle size of soil [51].

5.4 Physico-Chemical Properties of Soil

Physico-chemical properties of soil are important criteria for understanding the soil behavior. It includes the soil pH, soil temperature, electrical conductivity, and CEC.

5.4.1 Soil pH

The pH is a measure of acidity or basicity of soil. As per theory of dissociation, higher the degree of ionization, higher will be activity of ions in soil. pH may affect the soil properties by changing physical and chemical properties of soil. pH significantly affects plant and microbial growth [52]. According to literature, best suited pH range is 6.0–6.8 for most of crops because optimum solubility of plant nutrients is achieved at this pH range [52]. Aluminum and hydrogen are significant exchangeable cations in acidic medium. However, calcium and magnesium are important in basic medium of soil. In humid areas, soil is more acidic in nature because rainfall washes off alkaline cations and replaces them with H^+.

5.4.2 Soil Temperature

Soil temperature depends upon heat absorbed by surface of soil. It varies between −20°C and 65°C. Soil temperature influences physical, chemical, and biological properties of soil. Temperature of soil surface is generally different from inner layer of soil. Further, soil temperature not only varies in different seasons but also varies in day and night [53]. Generally, soil is warm in day time; however, it is cool at night. However, temperature does not change appreciably in deep layer of soil [54]. It has also been observed that biological activities of microorganism are more in warm soil as compared to cold soil [55].

5.4.3 Electrical Conductivity

It is the expression of presence of ions in soil. Health of soil can be determined by measuring electrical conductivity of soil. Fertile soil is having rich concentration of anions and cation in them. High amount of ionic content leads to high electrical conductivity in soil. It was found from literature that electrical conductivity in soil sediments is due to electro filtration, pH difference, ionic layer, and electro osmosis [56]. Soil salinity can be evaluated by measuring electrical conductivity of soil. As per literature, electrical conductivity of surface soil is very different from inner layer of soil [54]. According to soil scientists, clay is having high conductivity. However, silts have medium conductivity while sand is having low conductivity [54]. Further, strong correlation has been observed between electrical conductivity and particle size [56]. Thus measurement of electrical conductivity is significant tool for identifying various physico-chemical properties and pattern of crop yield [57]. Linear relationship has been observed between electrical conductivity and salinity of soil.

5.4.4 Cation Exchange Capacity

CEC is the capacity of soil to combine with cation in such a way that these cations cannot be removed easily but it can be exchanged by equal amount of other cations. CEC significantly affects the properties of soil by changing physical and chemical nature of soil [58]. CEC is major indicator for crop growth, soil fertility, and transport in soil [59]. The CEC is important factor which enhance the rate of adsorption of heavy metals and other pollutants in soil. Dutta *et al.* studied the effect of CEC on the adsorption of lead in soil [60]. They predicted adsorption and desorption behavior of soil. Literature also revealed the adsorption of tetracycline in soil and its correlation with CEC [61, 62]. CEC is also directly correlated with the amount of clay particles and its behavior in soil. The values of CEC fluctuate in soil and it can be in the range of 100–1000 kg/hectare. The correlation of organic matter, soil pH, and particle size with CEC has been studied in detail by soil scientists [63]. It was concluded from the research that sand, clay, and organic matter of soil greatly influence CEC of top layer of soil [64]. Literature also reveals that CEC of different types of soil are also different from one another [65].

5.5 Sorption Behavior of Soil

Surface chemistry of soil and its structure are mainly responsible for many industrial and environmental applications of soil. The chemistry of soil has a significant effect on the performance of soil as an adsorbent. The revolutionary work by many soil chemists regarding the role of soil as an adsorbent was done [66]. Chemical properties of soil affect directly or indirectly the sorption behavior of soil. As per literature, majority of minerals of soil are working very efficiently in removal of heavy metals and other adsorbents from aqeous medium [67]. Clay minerals are mainly made up of alumina and silica sheets. It has been seen in above sections that structure of minerals; exchangeable ions and ability of ions make the surface of soil highly porous which can be used as adsorbents. Few soil scientists studied and explored adsorptive characteristics of soil in natural and modified form [68]. The factors responsible for adsorption are pH, metal concentration, and dosage of amorphous clay and other minerals of soil. Natural clay play important role in removal of copper from wastewater. Most of clay particles are negatively charged so they adsorb metal ions very efficiently. As per literature, the adsorption mechanism was found between copper particles and functional groups of clay [69]. The removal of copper from wastewater is found to be 68.7% in the presence of clay as an adsorbent. Recently, removal of arsenic was studied from water sample with the help of five different concentration of clay. But, substantial removal could not be found for removal of arsenic [70]. Another study was reported for the removal of lead, cadmium and chromium from wastewater [71]. Clay was found to be good adsorbent for lead, cadmium, and chromium. Efficacy of heavy metals removal was observed to be greatly enhanced at high temperature [72, 73]. Successful chromium removal was found at 40°C. Results showed that dosage of clay for removal of heavy metals from water varied in the range of 0.743–1.422 mg/g [72, 73]. It was also found from literature that rate of removal of heavy metals was encouraging at high temperature and low pH value. Another study shows the opposite trend. It was observed that Cu (II) has been removed from water sample by natural clay [74]. In this case, optimum pH was found to be in range of 7.00–8.00. One literature shows the removal of Pb (II) ions by clay as an adsorbent using batch system. Results showed that best result was found at pH 7.00 [75].

It was found from literature that temperature and pH was two important factors for removals of heavy metals from water by clay as an adsorbent [72]. Adsorption was found to increase with increase in temperature to some extent and then starts decreasing with increasing temperature. Both endothermic and exothermic reactions have been observed during the process. Literature also showed that modified form of clay is also used efficiently for removal of heavy metals [76].

Modified form of montmorillonite is used successfully for removal of heavy metals from water. Hydrophilic form of montmorillonite cannot be used successfully as an adsorbent because water is already present in inner layer of silicate sheet. As per literature, montmorillonite can be modified to enhance the surface area so that it can become more porous [77]. Organo-modified montmorillonite was exploited for removal of copper. It could remove 99.2% copper at wide pH range of 3.0 to 8.0. It was also found from literature that organo-modified montmorillonite has been found to be effective in removal of Hg (II) and Pb (II) from aqueous medium [78]. The different findings show that organo-modified montmorillonite is consistent for removal of heavy metals. Almunium-modified montmorillonite has been exploited in removal of Cd (II) from water sample. But, result was not very encouraging at pH less than 6.0 [79]. It may be because of low concentration of H^+ ions on surface of montmorillonite at higher temperature which makes surface negatively charged. Recent study shows montmorillonite-illite was used for removal of Zn (II) and Cu (II) from industrial wastewater.

Montmorillonite-illite was collected from Karnataka for the study [80]. But, removal was not very encouraging. It may be because of interference of different components of montmorillonite-illite on adsorption sites.

Montmorillonite is also effective in its natural form in metal ion removal. Several literatures were reported about the use of montmorillonite in its natural form [81]. It is economical and potential adsorbent for removal of heavy metals. In recent work, Ca-montmorillonite and Na-montmorillonite were exploited to remove Ni (II) and Cu (II) from water sample [82].

Kaolinite has been used efficiently for adsorption of Cd (II) and Pb (II). It was observed from literature that adsorption of Cd (II) was done through endothermic reaction while Pb (II) adsorbed on kaolinite surface through exothermic reaction [83]. It was also found from literature that kaolinite can be used in both raw and modified form. Despite having low CEC, kaolinite is used efficiently in raw form for removal of heavy metals. Mechanism of removal was based on adsorption and ion exchange technique. Good

Table 5.6 Adsorption behavior of adsorbents for different adsorbates.

Adsorbates	**Adsorbents**	**Dosage**	**Temperature**	**Optimum pH**	**References**
Zn(II) Cd(II) Ni(II), Cu(II) Pb(II)	Montmorillonite	500 mg/L	25–26.8°C	5.0	11
Cs(II)	Iron pillared Montmorillonite	13.3 mg/L	15°C	7.3	86
Cu(II) Zn(II)	Ca- Montmorillonite	1N	NA	4.8	82
Cs (II)	Clay soil with Illite	4 mg/L	25°C	8.0	87
Pb(II) Cd(II) Ni(II)	Kaolinite	160 mg/L	30°C	7.0	84
Cs(II)	Soil clay with illite	4 mg/L	25°C	8.0	85
Hg(II)	Raw montmorillonite	100 mg/L	25°C	5	81
Cd (II) Pb(II)	Kaolinite	0.01–0.1 M	NA	4.0–7.0	83
Cr(II)	Amorphous clay	0.743–1.422 mg/g	40°C	7.0–8.0	73

results were obtained for removal of Cd (II), Ni (II), Pb (II), and Cu (II) at concentration range of 0.01–0.1M kaolinite solution [84].

Illite has been successfully used in removal of caesium from water. Comparative study for removal of caesium was done on montmorillonite, kaolinite and illite. Best result was obtained at pH 8.0 with illite as caesium showed best affinity with illite [85]. After comprehensive review on use of soil inorganic matter in water and wastewater treatment, results have been tabulated in Table 5.6.

5.6 Conclusion

Current studies in different field of soil chemistry have been extensively done here. Composition of soil greatly influences physico-chemical properties of soil. Soil inorganic matter comprises of primary and secondary minerals. Secondary minerals majorly consist of SiO_2, Al_2O_3 with small amount of Fe_2O_3, TiO_2, MgO, and K_2O. SOM also plays key role in various soil reaction and processes.

The concept of soil structure is having complex chemistry. From the chemistry point of view, soil structure is associated with different nutrient cycle and nutrient storage. It also controls storage of soil organic matter.

Soil color is found to be useful tool for determining physical and chemical properties of soil. Color of soil is mostly associated with presence of different elements in soil. Among sand, silt, and clay, texture of soil is mostly influenced by clay.

It has also been concluded that soil is having high sorption capacity and is very effective in removal of heavy metals from water and wastewater. A combination of soil minerals can also be used for complex industrial wastewater. Soil has many advantages in water chemistry. It is quite reasonable to believe that the soil inorganic matter in its natural and modified form is used efficiently in removal of heavy metals from wastewater. However, more quantitative study is required on this aspect to explore its further application in industrial wastewater.

References

1. Sparks, D.L. (Ed.), *Environmental Soil Chemistry*, Academic Press, San Diego, USA, 2003, https://doi.org/10.1016/B978-0-12-656446-4.X5000-2.
2. Khan, S., Jabin, S., Khan, J.A., Influence of some pesticides on mobility of some trace metals through soil amended with fly ash. *Pollut. Res.*, 19, 4, 607, 2000.

3. Khan, S., Khan, J.A., Bhardwaj, R.K., Jabin, S., Effect of some organic matter on the mobility of some trace metals through soil amended with fly ash. *J. Indian Chem. Soc.*, 77, 7, 326, 2000.
4. Sparks, D.L., Fundamentals of soil chemistry. *XXII International Symposium on homogenous catalysis*, Lisboa, Portugal, 2020, https://doi.org/10.1002/9781119300762.wsts0025.
5. Smita, K.T. and Ingole, S., A Review on role of physico-chemical properties in soil quality. *Chem. Sci. Rev. Lett.*, 4, 13, 57, 2015.
6. Zhou, Y., Liu, Q., Xu, P., Cheng, H., Molecular structure and decomposition kinetics of kaolinite/alkylamine interaction compounds. *Front. Chem.*, 6, 310, 2018.
7. Bhattacharya, K.G. and Gupta, S.S., Adsorption of heavy metals on natural and modified kaolinite and montmorillonite: A review. *Adv. Colloid Interface Sci.*, 140, 114, 2008.
8. Al-Essa, K. and Khalili, F., Heavy metals adsorption from aqueous solutions onto unmodified and modified Jordanian kaolinite clay: batch and column techniques. *Am. J. Appl. Chem.*, 6, 1, 25, 2018.
9. Clark, B.C., Arvidson, R.D., Gellert, R., Morris, R.V., Ming, D.W., Richter, L., Ruff, S.W., Michalski, J.R., Farrand, W.H., Yen, A., Herkenhoff, K.E., Evidence for montmorillonite or its compositional equivalent in Columbia Hills, Mars. *J. Geophys. Res. Solid Earth*, 112, 19, 2007.
10. de Menezes, L.R. and da Silva, E.O., The use of montrmorillonite clays as reinforcing fillers for dental adhesive. *Mater. Res.*, 19, 1, 236, https://doi.org/10.1590/1980-5373-MR-2015-0375, 2016.
11. Akpomie, K.G., Dawodu, F.A., Kayode, O., Adebowale, K.O., Mechanism on the sorption of heavy metals from binary-solution by a low cost montmorillonite and its desorption potential. *Alex. Eng. J.*, 54, 757, http://dx.doi.org/10.1016/j.aej.2015.03.0251, 2015.
12. Akpomie, K.G. and Dawodum, F.A., Acid-modified montmorillonite for sorption of heavy metals from automobile effluent. *J. Basic Appl. Sci.*, 5, 1, 1, https://doi.org/10.1016/j.bjbas.2016.01.003, 2016.
13. Uddin, F., Clay, nano-clay and montmorillonite minerals. *Metall. Mater. Trans. A*, 39, 12, 2804, 2008.
14. Uddin, M.K., A review on adsorption of heavy metals by clay minerals with special emphasis on past decades. *Chemical Eng. J.*, 308, 438, https://doi.org/10.1016/j.cej.2016.09.029, 2017.
15. Singh, M., Sarkar, B., Hussain, S., Churchman, J., Influence of physicochemical properties of soil clay fractions on the retention of dissolved organic carbon. *Environ. Geochem. Health*, 39, 6, 1335, 2017.
16. Alagoz, Z. and Yilmaz, E., Effects of different sources of organic matter on soil aggregate formation and stability: A laboratory study on Lithic Rhodoxeralf form of Turkey. *Soil Tillage Res.*, 103, 419, 2009.

17. Fageria, N.K., Role of soil organic matter in maintaining sustainability of cropping systems. *Commun. Soil Sci. Plant*, 43, 16, 2063, https://doi.org/10.1080/00103624.2012.697234, 2012.
18. Balesdent, J., Chenu, C., Balabane, M., Relationship of soil organic matter dynamics to physical protection and tillage. *Soil Tillage Res.*, 53, 215, 2000.
19. Ellerbrock, R.H., Gerek, H.H., Bachmann, J., Goebel, M.O., Composition of organic matter fractions for explaining wettability of three forest soils. *Soil. Sci. Soc. Am. J.*, 69, 57, 2005.
20. Orlov, D.S. and Sadovnikova, L.K., Soil organic matter and protective functions of humic substances in the biosphere, in: *Use of Humic substances to remediate Polluted Environments: From theory to practice*, I.V. Perminova and K. Hattfield (Eds.), pp. 32–52, Springer, Netherlands, 2005.
21. Hayers, M.H.B., Tseng, T.Y., Wang, M.K., Chemistry of soil organic matter. *Taiwan J. Sci.*, 22, 3, 215, 2007.
22. Hayes, M.H.B., Solvent systems for the isolation of organic components from soils. *Soil Sci. Soc. Am. J.*, 70, 986, 2006.
23. Singh, R.P., Parr, J.E., Stewart, B.A., *Advances in soil science*, vol. 13, p. 151, Springer Nature, USA, 1985.
24. Papadopoulos, A., Bird, N.R.A., Whitmore, A.P., Mooney, S.J., Does organic management lead to enhanced soil physical quality? *Geoderma*, 213, 435, https://doi.org/10.1061/j.geoderma.2013, 2014.
25. Manahan, S.E. (Ed.), *Fundamental of Environmental Chemistry*, CRC Press, Florida, USA, 2004.
26. Pierzynski, G.M., Sims, J.T., Vance, G.F. (Eds.), *Soils and Environmental Quality*, Taylor and Francis, USA, 2005, https://doi.org/10.1201/b12786.
27. Jury, W.A. and Horton, R. (Eds.), *Soil Physics*, John Wiley & Sons, New Jersey, USA, 2004.
28. Campbell, G.S. and Norman, J.M. (Eds.), *An Introduction to Environmental Biophysics*, Springer-Verlag, USA, 1998.
29. Askaer, L., Elberlingm, B., Glud, R.N., Kuhlm, M., Lauritsenm, F.R., Joensen, H.P., Soil heterogeneity effects on O_2 distribution and CH_4 emissions from wetlands: *In situ* and mesocosm studies with planar O_2 optodes and membrane inlet mass spectrometry. *Soil Biol. Biochem.*, 42, 2254, 2010.
30. Hers, I., Li, L., Hannam, S., Evaluation of soil gas sampling and analysis techniques at a former petrochemical plant site. *J. Environ. Technol.*, 25, 7, 847, 2004.
31. Winkel, A. and Borum, J., Use of sediment CO_2 by submersed rooted plants. *Ann. Bot.*, 103, 1015, https://doi.org/10.1093/aob/mcp036, 2009.
32. Heuscher, S.A., Brandt, C.C., Jardine, P.M., Using soil physical and chemical properties to estimate bulk density. *Soil Sci. Soc. Am. J.*, 69, 51, 2005.
33. Warkentin, B.R., Soil structure: A history from tilth to habitat. *Adv. Agron.*, 97, 239, 2008.

34. Brady, N. and Weill, R. (Eds.), *The nature and properties of soil*, Pearson Prentice Hall, New Jersey, 2008.
35. Lipiec, J. and Hatano, R., Quantification of compaction effects on soil physical properties and crop growth. *Geoderma*, 116, 107, 2003.
36. Gabreille, J.M., Lannoy, D., Koster, R.D., An updated treatment of soil texture and associated hydraulic properties in a global land modeling system. *J. Adv. Model. Earth Sy.*, 6, 4, 957, https://doi.org/10.1002/2014MS000330, 2014.
37. Park, E.J. and Smucker, A.J.M., Saturated hydraulic conductivity and porosity within macroaggregates modified by tillage. *Soil Sci. Soc. Am. J.*, 69, 38, 2005.
38. Six, J., A history of research on the link between micro aggregates, soil biota, and soil organic matter dynamics. *Soil Tillage Res.*, 79, 7, 2004.
39. Six, J., Elliott, E.T., Paustian, K., Soil structure and soil organic matter: A normalized stability index and the effect of mineralogy. *Soil Sci. Soc. Am. J.*, 64, 1042, 2000, 2004.
40. Bissonnais, Y., Aggregate stability and assessment of soil crustability and erodibility: Theory and Methodology. *Eur. J. Soil Sci.*, 47, 425, https://doi.org/10.1111/j.1365-2389.1996.tb01843.x, 1996.
41. Sanchez, M., Huertas, R., Melgosa, M., Color variation in standard soil color chart. *Aust. J. Soil Res.*, 43, 7, 827, 2005.
42. Barouchas, P.E. and Moustakas, N.K., Soil colour and spectral analysis employing linear regression models: Effect of organic matter. *Int. Agrophys.*, 18, 1, 10, 2004.
43. Bhadra, S.K. and Bhavanarayanam, M., Estimation of the influence of soil moisture on soil colour. *J. Plant Nutr. Soil Sci.*, 160, 4, 401, 1997.
44. Aitkenhead, M.J., Coull, M., Tower, W., Black, G.H.I., Prediction of soil characteristics and color using data from nation soils inventory of Scotland. *Geoderma*, 200, 99, http://dx.doi.org/10.1016/j.geoderma.2013.02.013, 2013.
45. Haidouti, C. and Massas, I., Distribution of iron and manganese oxides in Haploxeralfs and Rhodoxeralfs and their relation to the degree of soil development and soil colour. *J. Plant Nutr. Soil Sci.*, 161, 2, 141, https://doi.org/10.1002/jpln.1998.3581610209, 1998.
46. Moritsuka, N., Matsuoka, K., Katsura, K., Yanai, J., Soil color analysis for statistically estimating total carbon, total nitrogen and active iron contents in Japanese agricultural soils. *J. Soil Sci. Plant Nutr.*, 60, 4, 475, https://doi.org/10.1080/00380768.2014.906295, 2014.
47. Bernoux, M., Cerri, C., Arrouays, D., Jolivet, C., Bulk densities of Brazilian Amazon soils related to other soil properties. *Soil Sci. Soc. Am. J.*, 62, 743, 1998.
48. Hu, W., Shao, M.A., Si., B., Seasonal changes in surface bulk density and saturated hydraulic conductivity of natural landscapes. *Eur. J. Soil Sci.*, 63, 820, 2012.
49. Borkovec, M., Wu, Q., Degovics, G., Laggner, P., Sticher., H., Surface area and size distributions of soil particles. *Colloids Surf. A*, 73, 65, 1993.

50. Hwang, S., II and Powers, S.E., Using particle-size distribution models to estimate soil hydraulic properties. *Soil Sci. Soc. Am. J.*, 67, 1103, 2003.
51. Lucian, C., Characterization of Particle Size Distribution in Expansive Soils using Logarithmic Density Distribution. *Int. J. Eng. Trends Technol.*, 49, 3, 161, 2017.
52. Andersson, S. and Nilsson, S., II, Influence of pH and temperature on microbial activity, substrate availability of soil-solution bacteria and leaching of dissolved organic carbon in a mor humus. *Soil Biol. Biochem.*, 33, 1181, 2001.
53. Solanki, H.A. and Chavda, N.H., Physico-chemical analysis with reference to seasonal changes in soils of Victoria park reserve forest, Bhavnagar (Gujarat). *Life Sci. Leafl.*, 8, 62, 2012.
54. Wang, H.M., Wang, W.J., Chen, H.F., Zhang, Z.H., Mao, Z.J., Zu, Y.G., Temporal changes of soil physic-chemical properties at different soil depths during larch afforestation by multivariate analysis of covariance. *Ecol. Evol.*, 4, 7, 1039, 2014.
55. Onwuka, B. and Mang, B., Effect of soil temperature on some soil properties and plant growth. *Adv. Plants Agric. Res.*, 8, 1, 34, 2018.
56. Aini, I.N., Ezrin, M.H., Aimrun, W., Relationship between soil electrical conductivity and pH value of Jawa series in oil palm plantation. *Agric. Agric. Sci. Procedia*, 2, 199, 2014.
57. Kekane, S.S., Chavan, R.P., Shinde, D.N., Patil, C.L., Sagar, S.S., A review on physico-chemical properties of soil. *Int. J. Chem. Stud.*, 3, 4, 29, 2015.
58. Zhao, X., Determination and analysis of cation exchange capacity of agricultural soil in xiangyang city. *J. Environ. Dev.*, 28, 1, 53, 2016.
59. Revil, A. and Leroy, P., Governing equations for ionic transport in porous shales. *J. Geophys. Res.*, 109, 3208, 2004.
60. Dutta, S.K., Singh, D., Sood, A., Effect of soil chemical and physical properties on sorption and desorption behavior of lead in different soils of India. *Soil Sediment Contam.*, 20, 3, 249, 2011.
61. Teixido, M., Granados, M., Prat, M.D., Sorption of tetracyclines onto natural soils: data analysis and prediction. *Environ. Sci. Pollut. Res.*, 19, 8, 3087, https://doi.org/10.1021/es06231, 2012.
62. Moorman, T.B., Jayachandran, K., Reungsang, A., Adsorption and desorption of atrazine in soils and subsurface sediments. *Soil Sci.*, 166, 921, 2011.
63. Kathrin, O.R., Cauley, A., Jones, C., Soil pH and organic matter, in: *Nutrient Management*, Module-8, March 2017.
64. Ghorbani, H., Kashi, H., Moghadas, N.H., Estimation of soil cation exchange capacity using multiple regression artificial neural networks and adaptive neuro-fuzzy inference System models in Golestan Province, Iran. *Commun. Soil Sci. Plant*, 46, 6, 763, https://doi.org/10.1080/00103624.2015.1006367, 2015.

65. Środon, J. and McCarty, D.K., Surface area and layer charge of smectite from CEC and EGME=H_2O-Retention measurements. *Clay Miner.*, 56, 2, 155, 2008.
66. Sangiumsak, N. and Punrattanasin, P., Adsorption behaviour of heavy metals on various soil. *Pol. J. Environ. Stud.*, 23, 3, 853, 2014.
67. Etci, O., Bektas, N., Oncle, M.S., Single and binary adsorption of lead and cadmium ions from aqueous solution using the clay mineral. *Environ. Earth Sci.*, 61, 231, 2010.
68. Chantawong, V., Harvey, N.W., Bashkin, V.N., Comparison of heavy metal adsorptions of Thai kaolin and ball clay. *Water Air Soil Pollut.*, 148, 111, 2003.
69. Srinivasan, R., Advances in application of natural clay and its composites in removal of biological, organic, and inorganic contaminants from drinking water. *Adv. Mater. Sci. Eng.*, 2011, 17, 2011.
70. Bentahar, Y., Hurel, C., Draoui, K., Khairoun, S., Marmier, N., Adsorptive properties of Moroccan clays for the removal of arsenic(V) from aqueous solution. *Appl. Clay Sci.*, 119, 385, 2016.
71. Khan, T.A. and Singh, V.V., Removal of cadmium (II), lead (II), and chromium (VI) ions from aqueous solution using clay. *Toxicol. Environ. Chem.*, 92, 1435, 2010.
72. Sreedharam, V. and Nidheesh, P.V., Removal of chromium and iron from real textile wastewater by sorption on soil. *J. Hazard Toxic Radioact. Waste*, 21, 4, 2017.
73. Weng, C.H., Sharma, Y.C., Chu, S.H., Adsorption of Cr (VI) from aqueous solutions by spent activated clay. *J. Hazard. Mater.*, 155, 65–75, 2008.
74. Zacaroni, L.M., Magriotis, Z.M., Cardoso, M.G., Santiago, W.D., Mendonça, J.G., Vieira, S., Nelson, D.L., Natural clay and commercial activated charcoal: Properties and application for the removal of copper from cachaça. *Food Control*, 47, 536–544, 2015.
75. Eloussaief, M. and Benzina, M., Efficiency of natural and acid-activated clays in the removal of Pb(II) from aqueous solutions. *J. Hazard. Mater.*, 178, 753, 2010.
76. Bosco, S.M.D., Jimenez, R.S., Vignado, C., Fontana, J., Geraldo, B., Figueiredo, F.C.A., Mandelli, D., Carvalho, W.A., Removal of Mn (II) and Cd (II) from wastewaters by natural and modified clays. *Adsorption*, 12, 133, 2006.
77. Sherbini, K. and Hassanien, M.M., Study of organically-modified montmorillonite clay for the removal of copper (II). *J. Hazard. Mater.*, 184, 654, 2010.
78. Cruz-Guzmana, M., Celis, R., Hermosina, M.C., Koskinen, W.C., Nater, E.A., Cornejo, J., Heavy metal adsorption by Montmorillonites modified with natural organic cations. *Soil Sci. Soc. Am.*, 70, 215, 2006.
79. Yu, R., Wang, S., Wang, D., Ke, J., Xing, X., Kumada, N., Kinomura, N., Removal of Cd^{2+} from aqueous solution with carbon modified aluminum-pillared montmorillonite. *Catal. Today*, 139, 135, 2008.

80. Oubagaranadin, J.U.K., Murthy, Z.V.P., Mallapur, V.P., Removal of Cu (II) and Zn (II) from industrial wastewater by acid-activated montmorillonite-illite type of clay. *C.R. Chim.*, 13, 1359, 2010.
81. Tran, L., Wu, P., Zhu, Y., Liu, S., Zhu, N., Comparative study of Hg (II) adsorption by thiol and hydroxyl containing montmorillonites. *Appl. Surf. Sci.*, 356, 91, 2015.
82. De-Pablo, L., Chavez, M.L., Abatal, M., Adsorption of heavy metals in acid to alkaline environments by montmorillonite and Ca-montmorillonite. *Chem. Eng. J.*, 171, 1276, 2011.
83. Gupta, S.S. and Bhattacharya, K.G., Treatment of water contaminated with Pb (II) and Cd (II) by adsorption on Kaolinite and Montmorillonite and their acid activated forms. *Indian J. Chem. Technol.*, 16, 457, 2009.
84. Rao, R.A.K. and Kashifuddin, M., Adsorption studies of Cd (II) on ball clay: Comparison with other natural clays. *Arab. J. Chem.*, 9, 1233, 2016.
85. Giannakopoulou, F., Haidouti, C., Chronopoulou, A., Gasparatos, D., Sorption behavior of cesium on various soils under different pH levels. *J. Hazard. Mater.*, 149, 553, 2007.
86. Ararem, A., Bouras, O., Arbaoui, F., Adsorption of caesium from aqueous solution on binary mixture of iron pillared layered montmorillonite and goethite. *Chem. Eng. J.*, 172, 1, 230, 2011.
87. Giannakopoulou, F., Haidouti, C., Chronopoulou, A., Gasparatos, D., Sorption behavior of cesium on various soils under different pH levels. *J. Hazard. Mater.*, 149, 553, 2007.

6

Soil and Their Contaminants

Anupama Rajput

Department of Chemistry, MRIIRS, Faridabad, India

Abstract

Soil is known to be combination of inorganic and organic solids, air, and water. It is a really necessary constituent of the lithosphere. Soil may be a complicated physico-organic structure pertaining water, mineral salts, and supplements and broke up oxygen. Soil is capable of holding water, providing nutrients and medium for plant growth and supporting a good vary of organic phenomenon communities and a modifier of the atmosphere. Soil is additionally a vital environmental agent acting as a "filter" for liquid and solid inputs as well as rain, municipal waste, pesticides, and alternative chemicals. Soil chemistry contains the chemical reactions and processes between these constituents and principally emphases on considering the destiny of pollutants and nutrients inside soils. Information of soil chemistry permits inventors to observer and management and predicts the consequences of contaminants within the setting. Chemical information shared with concerns from the planet sciences, physics, and biology area unit desired to grasp, stop, and repair environmental problems with soils. Present chapter provides a comprehensive edaphology of soil and environmental reactions between soils and alternative natural material and significant metal.

Keywords: Lithosphere, nutrient, mineral salts, pollutants, contaminants, heavy metal, soils

6.1 Introduction

6.1.1 Component of Soil

Soil is an especially imperative ingredient of the lithosphere. It is a film of the globe shell with existing organisms and their product of decompose.

Email: anupama.fet@mriu.edu.in

Inamuddin, Mohd Imran Ahamed, Rajender Boddula and Tariq Altalhi (eds.) Applied Soil Chemistry, (105–122) © 2021 Scrivener Publishing LLC

It is a composite contain stream, sandstone salts, nutrients, and dissolved O_2. The basic composition of earth crust is the following elements present in soil by weight (shown in Table 6.1).

Soil is an intricate natural system comprise of natural and inert material including water and gas es in changeable proportions. The major constituents of soil are mineral matter, organic matter, and soil water. Soils consist of soluble salts such as chlorides, nitrates, sulfate carbonates, and bicarbonates of sodium, calcium, and magnesium.

Phosphorus happens inside the soil also in each natural and inorganic type [1]. The inorganic type is a lot of necessary for the harvest nourishment [2, 3]; nearly all of the P is fascinated by the plants as HPO_4 and H_2PO_4 ions or miscible crude phosphates [4]. The accessibility of phosphorus in the dirt is amazingly flexible in light of the fact that it relies upon the mineral soil creation, natural materials and its pace of deterioration, local atmosphere, and, along these lines, the morphological properties of soil. The synthetic component inadequacy can bring about staining of more established leaves [5, 6]. Soil surface also impacts improvement frameworks thus in a roundabout way soil breath. By and large, root development is more slow in the dirt of coarser surface

Table 6.1 Elementary composition of the Earth's crust.

Element present	Percentage by weight
O_2	48%–49%
Al_3	7.0%–7.02%
Si	25%–26%
Fe	4.0%–4.12%
Ca	3.0%–3.18%
K	2.0%–2.30%
Na	2.0%–2.33%
Mg	2.0%–2.11%
H_2	0.9%–0.97%
Ti	0.0%–0.41%
Cl_2	0.0%–0.20%

(increasingly sandy) than of better surface (less sandy) on account of lower ripeness, lower unsaturated pressure–driven physical wonder, and lower water stockpiling capacity. High root biomass and creation end in high paces of root breath and along these lines the related microbial breath inside the rhizosphere [7].

6.1.2 Major Types of Soil in India

Soil occurs in different part of our country [19] (shown in Figure 6.1).

6.1.3 The Various Factors Occurs in Soil Formation

The relief features parent material climate vegetation & other life forms as well as time apart from human activities are the major factors responsible for the formation of soil (shown in Figure 6.2).

6.1.4 Types of Soil Textures

6.1.4.1 Clayer Soils

The particle size is less than 0.002 mm. They have very small interspaces and hence inhibit free circulation of water and air. This type of soil is unsuitable for growth of plants.

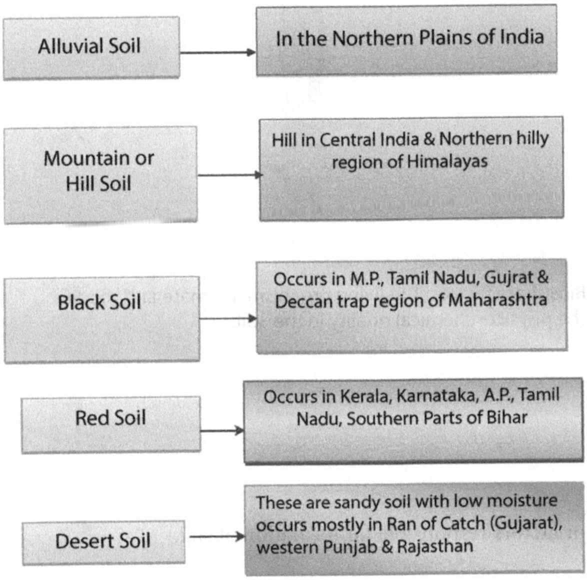

Figure 6.1 Primary composition of the Earth's crust.

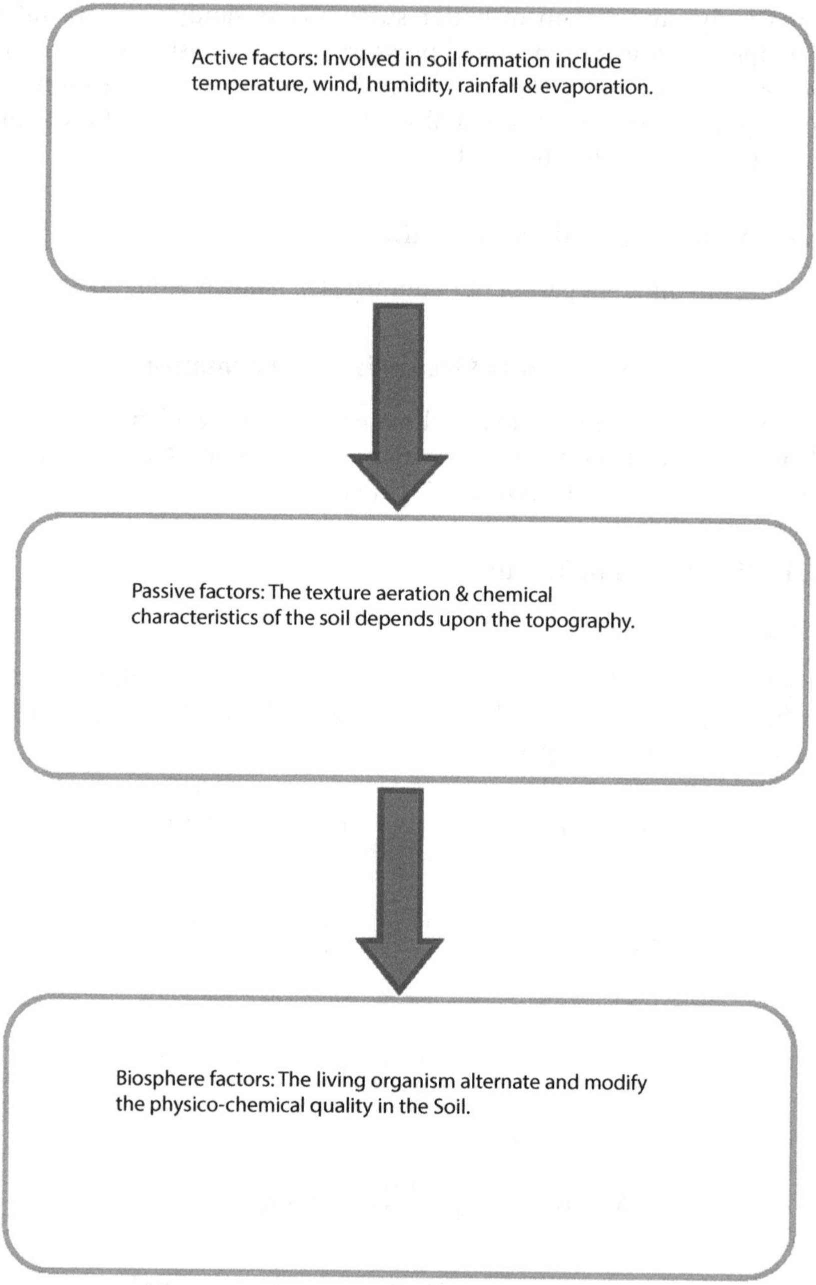

Figure 6.2 Major factors responsible for formation of soil.

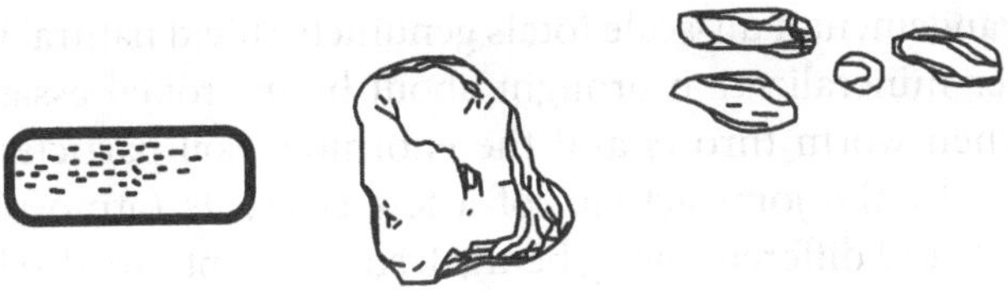

Figure 6.3 Types of soil texture.

6.1.4.2 Sandy Soils

It consists of loose and dry sand particles; they have poor water holding capacity and less nutrients content.

6.1.4.3 Slit Soils

It contains sand, clay, and slit. This type of soil is the best for plant growth depending upon the proportions of sand, clay, and slit and is not suitable for plant growth (shown in Figure 6.3).

6.2 Soil Organic Matter (SOM)

The organic matter content of a soil is generally 2%–5% partially decomposed organic matter, i.e., humus plays a very vital position in maintaining the porosity of the soil and its nutrient and water retaining capacity and fertility of soil. Humus also helps to bind the soil erosion.

The top layer of soil consists of organic matter such as plants and multiplicity of organisms such as bacteria, algae, fungi, protozoa, arthropods, etc. Soil assumes an indispensable job in deciding the value and organization of the biosphere which creates over it. Soil provides mechanical support to the plants soil—contains organ trophic bacteria, nitrogen fixing bacteria, fungi protozoans, and other microbes, which helps in decomposition and mineralization of natural substance and rejuvenation of nutrients. The colloidal component of soils which comprises of clay micelles and human particles (less than 0.002 mm) lightly absorbs a number of nutrient ions and while maintaining proper PH."

Soil natural issue will, in general, increment as the clay content increments. This expansion relies upon two components. To start with, bonds between the outside of mud particles and natural issues impede the decay procedure. Second, soils with higher dirt substance increment the potential for total arrangement. Full-scale totals genuinely shield natural issue atoms from further mineralization brought about by microbial assault [8]. For instance, when worm throws and the enormous soil particles they contain are part by the joint activity of a few elements (atmosphere, plant development, and different living beings), supplements are discharged and made accessible to different segments of soil miniaturized scale creatures. Under comparative atmosphere conditions, the natural issue content in fine-finished (clayey) soils is two to multiple times that of coarse-finished (sandy) soils [9].

Ca transpires in the soil furthermore as dissolvable $Ca2^{+}$ on the base multifarious or as permitted calcium carbonate (CaCO3) [10].

Magnesium is also an important constituent of soil. It likewise initiates the number of plant catalysts. In the event that the dirt has a lack of Mg, at that point, the plant developed in such soil will get light yellow and afterward turns dark-colored. Potassium is present in the dirt in various structures [11].

6.2.1 Soil Chemical Reaction

Soil pH is a significant element constraint as it helps in ensuring availability of plant essential nutrients [12] the pH of soil was varied from 4.49 to 7.58 with a mean value of 5.98. This indicates moderately acidic soil reaction (pH). The pH of soil samples was found to be 41.98% of sample showed moderately acidic, 20.99% samples were slightly acidic, 11.11% samples were very acidic, while 16.05% samples were nearly neutral and only 9.88% samples were slightly alkaline in nature. The soils are acidic might be as a result of the leaching of basic cation or due to incessant uptake by crops grown on the field [13]. The soil acidity implied that nutrients are likely to be available or unavailable.

6.2.2 Sality and Acidity

The quality, toxicity, and limits in soil pH scale (acidic or alkaline) end in underprivileged biomass making and so in reduced additions of organic refer the soil. For instance, pH scale affects humus arrangement in two ways: putrefaction and biomass creation. In powerfully acid or extremely alkaline soils, the growing conditions for micro-organisms square measure poor, leading to low levels of biological oxidation of organic matter (Primavesi, 1984) [14].

Soil acidity conjointly manipulates the contribution of plant nutrients and so normalizes that in some way to biomass production and therefore the accessible food for soil collection. Fungi square calculates less receptive than bacterium to acid soil situation [14].

6.3 Contaminants in Soil

Soil defilement is obviously fetched by automatic movement, agricultural synthetics, or improper removal of ravage. The major recognized synthetic substances concerned square measure fossil oil hydrocarbons, poly nuclear fragrant hydrocarbons [for example, hydrocarbon, and benzo(a)pyrene], solvents, pesticides, lead, and option critical metals. Defilement is corresponded with the level of mechanical venture and power of concoction substance. Soil sullying stems basically from well-being dangers, from direct contact with the defiled soil, fumes' from the contaminants, or from auxiliary pollution of water gives inside and fundamental to the dirt. (Panagos *et al.*, 2013) [15]. The most widely recognized synthetics concerned square measure fossil oil hydrocarbons, solvents, pesticides, lead, and option huge metals. Any movement that winds up in elective sorts of soil corruption (disintegration, compaction, and so forth) may in a roundabout way break down the tainting impacts right now turns into a ton of dreary. Micro plastics square measure terribly tiny items of plastic that contaminate the atmosphere likewise as soil (Blair *et al.*, 2016) [16]. They enter natural eco system sand contaminants them (Collignon *et al.*, 2014) [17]. The major sources of metallic contamination of soil represent mining, smelting, sludge, fertilizers, pesticides, etc. Metals like Cd, Hg, Ni, Mo, and Cr are harmful to plant and animal life. Indiscriminate marketing of industrial wastes ends up in the natural action and flowing of harmful substances in to the soil and pollution of H_2O. Fly debris created from warm force plants, modern waste released into streams or drop into the encompassing area, mining squanders, non-biodegradable natural poisons, mechanical slime like vent gas, and noteworthy metal slop, and so forth, cause genuine water and soil contamination. Tainted or defiled soil uncomplicatedly manipulates human well-being through direct contact with soil or by means of inward breath of soil contaminants that have disintegrated; without a doubt, greater dangers square measure show by the penetration of soil pollution into groundwater springs utilized for human utilization, by and large in territories clearly such a great amount off from any evident inventory of higher than ground sullying. This will, in general, lead to the occasion of contamination related ailments. Well-being outcomes from presentation to soil pollution shift extraordinarily wagering on squander material sort, the pathway of assault and powerlessness of the uncovered populace. Interminable presentation to chromium, lead, and elective metals, oil, solvents, and a lot of concoction and weed executioner details might be the disease, will cause non-heritable clutters, or will cause elective incessant well-being conditions. Modern or stunning convergences of present substances, similar to nitrate and smelling salts identified with eutherian fertilizer from farming activities, have also been known as well-being risks in soil and groundwater. Well-being impacts like cerebral pain, queasiness, weariness, eye disturbance, and blossoming for the higher than referred to and elective synthetic compounds. At save measurements, an outsized scope of soil contaminants will cause demise by a presentation by means of direct contact, inward breath, or take-up of contaminants in groundwater defiled through soil (Michael *et al.*, 1973) [18].

6.3.1 Pesticides

As per the reports of World Health Organization about 50,000 people in developing countries are poisoned and about 5,000 peoples die because of improper use of pesticides and other chemicals in modern agricultural techniques.

Pesticides create probable vulnerability to animals, humans, and water living. They furthermore source injurious consequence on soil productiveness and crop yield. Pesticides imposed to crops are retained in the soil in considerable quantities. They entered in to cyclic environmental processes such as absorption by soil and leaching by water.

Pesticides together with herbicides, fungicides, insecticides, and rodenticides are persistent contaminant. They enter to the food chain and pose serious health hazards. Some of the pesticieds are more dangerous to soil [19].

6.3.2 Some of the Adverse Effects of Pesticides

Pesticides are hazards to animals, humans and aquatic life. They also cause deleterious effects on soil fertility and crop productivity. Pesticides may enter the food chain and pose serious health hazards. Exessive use of synthetic pesticides may lead to defoliation of forest and adverse effect on fauna and flora (shown in Figure 6.4).

6.4 Pollution of Soil

The major sources of contamination of soils include mining, sludge, fertilizers, pesticides etc. Indiscriminate dumping of industrial waste & municipal wastes leads to the leaching in to the soil and can cause pollution of ground water (shown in Figure 6.5).

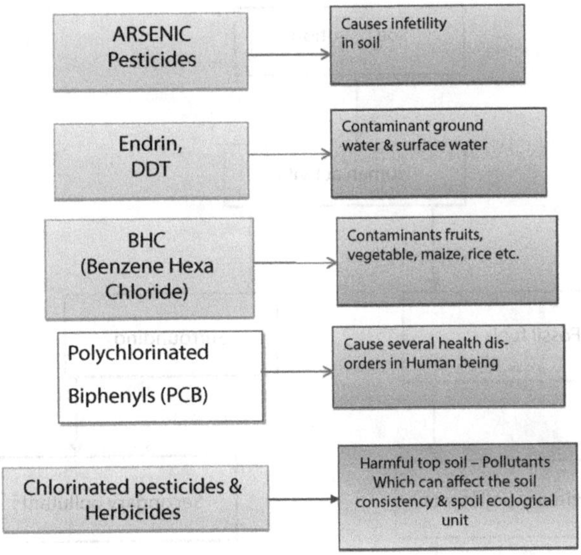

Figure 6.4 Calamitous effects of pesticides.

6.4.1 Effects of Modern Agriculture

Synthetic fertilizers are employed to increase the soil fertility and crop productivity. Excessive use of chemical fertilizers may cause undesirable effects. Contamination due to nitrate exceeding 90 PPM in drinking water may lead to diarrhea, blue Jaundice (Cyanosis) in children.

Vegetation growth in nitrate rich soils may exert toxic effects in cattle. Excessive use of chemical fertilizers may enter the water bodies, which causes eutrophication.

Excessive use of potassium fertilizers in soils may reduce the quantities of valuable ascorbic acid (vitamin C) and carotene in fruits and vegetable growth in such soil. The large-sized fruits and vegetables growth in highly fertilize soil and may be more attack by pests and insects [19].

6.4.2 Effects of Chemicals

Soil contagion as a piece of land degradation is brought about by the nearness of xenobiotics (human-made) synthetic concoctions or various adjustments inside the regular soil setting. It, for the most part, conveys about by mechanical action, unindustrialized synthetic composites, or inappropriate removal of ravage. The chief basic synthetic compounds concerned are raw petroleum hydrocarbons, polynuclear fragrant hydrocarbons, [for example, hydrocarbon and benzo(a)pyrene], solvents, pesticides, lead, and

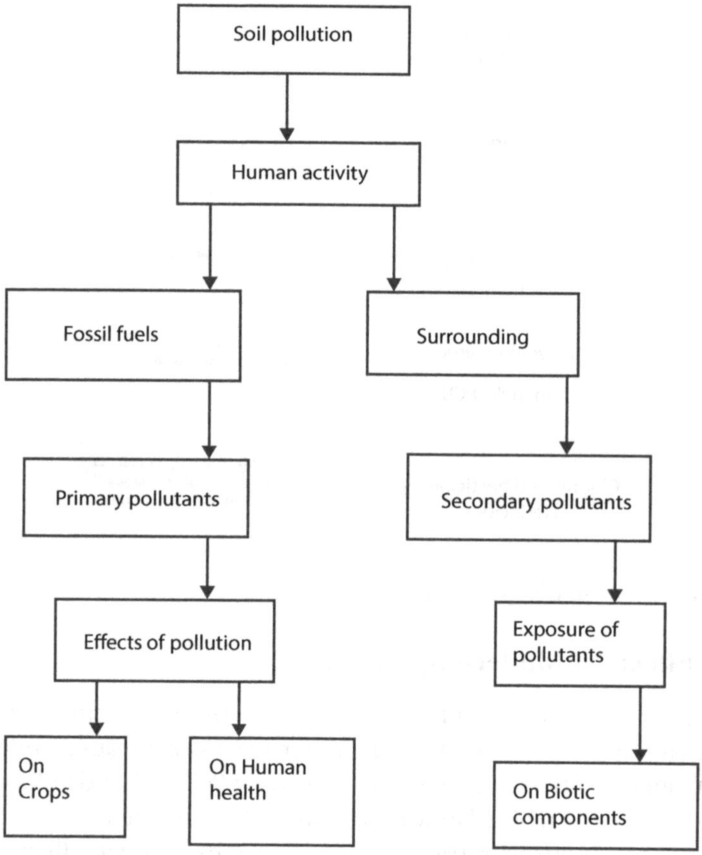

Figure 6.5 Major sources of soil pollution.

diverse huge metals. Sullying is identified with the level of industrialization and the power of compound substances. The need over soil pollution stems principally from well-being dangers, from direct contact with the tainted soil, fumes from the contaminants, or from auxiliary defilement of water given inside, furthermore, hidden the soil [20]. Mapping of tainted soil destinations and furthermore the following cleanups are time serious and high-ticket assignments, requiring inside and out measures of earth science, hydrology, science, PC demonstrating aptitudes, and GIS in environmental contamination, in like manner as an energy about the historical backdrop of business chemistry [21].

Major reason of soil contamination are a unit as pursue: industrialized movements, particularly in view of the fact that the number of mining and producing has augmented; agricultural activities, insect killers/pesticides, and fertilizers that area unit packed with chemicals that do not seem to be absolutely degradable in nature and area unit wide used round the world; waste disposal, wherever there is additionally an oversized quantity of commercial and community ravage or waste that is drop directly into landfills with none treatment; and incidental oil slicks, any place oil holes will occur all through capacity and transport of synthetic concoctions.

Primary effects of soil contamination region unit sway on the well-being of people, impact on the development of plants, decreased soil ripeness, and cyano genic mud. Through expanding our comprehension and improvement of inventive procedures to research and treat sullied soils, researchers and designers will assume a fundamental job in conveyance models and innovations to deal with the environmental factors contamination issue efficiently [22].

Skala *et al.* [23] considered the commitment of organo chlorine pesticides, polycyclic sweet-smelling hydrocarbons, and polychlorinated biphenyls to human well-being dangers in floodplains. Well-being hazard evaluation was what is more consideration of the examination by Retamel-Salgado *et al.* [24]. All through this case, the well-being hazard uncovered by the utilization of maize developing in three soils with contrastive qualities unclean with number forty-eight was assessed.

Yang *et al.* [25] utilized and treated the soil squander item ooze as a partner added substance to relate urban nursery soil. Absolute fixations, non-residual parts, and bio-accessibility of metal, Cu, Pb, and number thirty expanded once the expansion of treated the soil squander item slime to the dirt.

6.5 Chemistry of Saline Soil

An elevated level of salts during dirt will curtail the molecule cooperation in light of metallic component particles at the same time, for palatable plant execution, the salt fixation inside the dirt goals ought to be controlled so yield or field creation is not unfavorably influenced. A review of the essential powers worried in molecule association is given as a foundation to an idea of aftereffects of metallic component particles on the structure of soil and consequently the ways that inside which the impact of metallic component is reduced.

It is especially related that the consequence of metallic component particles on the expanding properties of dirt is along these lines articulated that even in soils of daylight surface, similar to sandy soil, awfully stamped changes in soil porosity is additionally concerned if the dirt has partner significant replaceable metallic component share.

Soil includes ample impartial soluble salts to unfavorably have an effect on the expansion of most crop or plant life. The soluble salts are a unit in the main common salt and sodium sulfate. However, saline soils conjointly have considerable quantities of chlorides and sulfates of Ca and metallic element.

The expenses of saltiness are

- detrimental impacts on plant development and yield;
- damage to the foundation (streets, blocks, consumption of channels and links);
- reduction of water quality for clients, alluviation issues, duplicated regular activity of metals,
- especially copper, cadmium, manganese, and zinc;
- soil disintegration at last, when crops zone unit excessively capably stricken by the measures of salts;
- soil saltiness is decreased by normal activity dissolvable salts out of the dirt with abundant water system water;
- soil saltiness, the executives include water level administration and flushing together with tile
- voidance or another sort of underground voidance [26];
- elevated levels of soil saltiness are endured if salt-tolerant plants territory unit grown-up.

Touchy yields lose their life as of now in marginally saline soils, most harvests territory unit adversely stricken by (modestly) saline soils, and exclusively saltiness safe yields flourish in seriously saline soils. The University of Wyoming [27] and along these lines the Government of Alberta reports data on the salt resilience of plants.

Field data in inundated grounds, underneath ranchers' conditions, are rare, especially in creating nations. In any case, some on-ranch reviews are made in Egypt [28], India [29], and West Pakistan [30]. Salt focused on the soils zone unit are well known to stifle the extension of plants 2 [31]. Plants in their normal setting region unit are colonized each by endo cell and energize thing microorganisms [32]. Rhizosphere microorganisms, outstandingly supportive bacterium and parasites, will improve plant execution underneath stress conditions and, therefore, upgrade yield each legitimately and indirectly [33].

Several plant growth progress to rhizobacteria (PGPR) could be appropriate to an instant provocation on plant growth and advancement by furnishing plants with the secured component, phytohormones, "iron that has been sequestered by microorganism siderophores, and solvent phosphate (Hayat *et al.*, 2010) [34]. Others do that in a roundabout way by ensuring the plant against soil-borne disease, a large portion of that region unit brought about by morbific organisms (Lutgtenberg and Kamilova, 2009) [35]. The matter of soil salinization might be afflicted for agrarian efficiency around the world.

Yields develop on saline soils suffer on collaborator proceedings of high dispersion stress, natural procedure issue and toxicities, poor soil physical conditions, and diminished harvest efficiency. This survey centers on the enhancement of effectiveness underneath focused on conditions and increased opposition of plants against saltiness worry by utilization of plant growth and advancing microorganisms.

Plants delicate to those segments are likewise influenced at similarly low salt focuses if the dirt contains enough of the hepatotoxic part. As a consequence of a few salts in plant supplements, high salt levels inside the dirt will agitate the supplement balance inside the plant or meddle with the take-up of certain supplements [36]. Salinity conjointly influences substance change mainly through a markdown in leaf space, chlorophyll content, and stomatal electrical marvel and, to a lesser degree, through a diminishing in photograph framework II efficiency [37].

6.6 The Effect of Salinity in Plant Growth

The venomous property is brought about by high saltiness on plant development, including plant hereditary designing [38] and, as of late, the utilization of plant development advancing microbes (PGPB). The role of microorganisms in plant development, supplement of the executives, and unwellness of the board is standard and entrenched. These supportive microorganisms colonize the rhizosphere/endorhizosphere of plants and advance the development of the plants through shifted immediate and roundabout systems [39, 40]. Past investigations prescribe that the use of PGPB has gotten promising different to lighten plant pressure brought about by saltiness [41], and consequently, the job of microorganisms inside the administration of natural marvel and abiotic stresses is picking up significance.

Saltiness also influences synthetic procedure primarily through a markdown in leaf space, shade content, and stomatal electrical wonder and, to a lesser degree, through a reduction in photosystem II potency [42]. Salinity antagonistically influences procreative advancement by possessing microsporogenesis and conceptive structure fiber prolongation, upgrading customized passing in some tissue assortments, ovule fetus removal, and senescence of inseminated.

The saline development medium causes a few unfavorable impacts on plant development, because of an intermittent dispersion capability of soil goals (osmotic pressure), explicit molecule impacts (salt pressure), natural procedure lopsided characteristics, or a blend of those variables [43]. This causes antagonistic consequences for plant development and advancement at physiological and natural science levels [44] and at the atomic level [45].

6.7 Conclusion

A perfect property of horticultural framework is one that keeps up and improves human well-being, preferences makers and clients each financially, and profoundly ensures the environment. Associate in nursing creates enough nourishment. One in all the principal fundamental limitations to agrarian creation on the planet is abiotic stress conditions winning inside the climate. Plant-related microorganisms will assume a crucial job in giving protection from abiotic stresses. These life forms may exemplify rhizoplane, rhizosphere, and endophytic microorganism and ward organisms and work through a spread of components like activating dissemination reaction, giving development hormones and supplements, acting as bio the executives specialists and enlistment of novel qualities in plants. The occasion of pressure tolerant yields assortments through quality joining and plant rearing is significant anyway an extended drawn and costly strategy, while microorganism immunization to mitigate worries in plants might be a great deal of cost viable earth inviting decision that might be open during a shorter time period. Taking this leads accessible, joint future analysis is required during this space, notably on field analysis and application of potential organisms as bio fertilizers in stressed soil.

References

1. Pettygrove, G.S. and Bay, I., *Crop nutrient harvest removal*, University of California Cooperative Extension. Manure Technical Bulletin Series, http://manuremanagement.ucdavis.edu, 2009.
2. Wan, J.H.C. and Wong, M.H., Effects of earthworm activity and P-solubilising bacteria on P availability in soil. *J. Plant Nutr. Soil Sci.*, 187, 209–213, 2004.
3. Niklas, K.J., Owens, T., Reich, P.B., Cobb, E.D., Nitrogen/phosphorus leaf stoichiometry and the scaling of plant growth, https://doi.org/10.1111/j.1461-0248.2005.00759.x, 2005.

4. Reynolds, C.M., Wolf, D.C., Armbruster, J.A., Factors related to urea hydrolysis in soil. *Soil Sci. Soc. Am. J.*, 49, 104, 1985.
5. Van Duren, I.C. and Pegtel, D.M., Nutrient limitations in wet, drained and rewetted fen meadows: evaluation of methods and results. 220, 1, 35, 2000.
6. Oladokun, M.A.O., Vegetative propagation of kola nut trees. *Ife J. Agric. For.*, 1990.
7. Högberg, P., Nordgren, A., Göran, I., Oecologi, A., Carbon Allocation between Tree Root Growth and Root Respiration in Boreal. Pine Forest. 132, 4, 579, 2002.
8. Rice, C.W., Organic matter and nutrient dynamics, in: *Encyclopedia of soil science*, p. 925, Marcel Dekker Inc., New York, USA, 2002.
9. Prasad, R. and Power, J.F., *Soil fertility management for sustainable agriculture*, p. 356, Lewis Publishers., New York, USA, 1997.
10. Sandaa, R.A., Torsvik, V., Enger, O., Influence of long-term heavy-metal contamination on microbial communities in soil. *Soil Biol. Biochem.*, 33, 3, 287–295, 2001.
11. Van Dam, J.C., Groenendijk, P., Hendriks, R.F.A., Kroes, J.G., Advances of modeling water flow in variably saturated soils with SWAP. *Vadose Zone J.*, 7, 640–653, 2008.
12. Deshmukh, K.K., Studies on chemical characteristics and classification of soils from Sangamner area, Ahmadnagar District, Maharastra. Rasayan. *J. Chem.*, 5, 1, 74, 2012.
13. Brady, N.C. and Weil, R.R., *The nature and properties of soils*, 13th edition, Pearson Education, New Jersey, 2002.
14. Primavesi, A., *Manejo ecológico del suelo. La agricultura en regiones tropicales*, 5ta Edición, p. 499, El Ateneo, Rio de Janeiro, Brazil, 1984.
15. Panagos, P., Van Liedekerke, M., Yigini, Y., Montanarella, L., Contaminated Sites in Europe: Review of the Current Situation Based on Data Collected through a European Network. *J. Environ. Public Health*, 2013.
16. Crawford, B.C. and Quinn, B., *Micro Plastic Pollutants*, 1st ed, Elsevier Science, 2016.
17. Collignon, A., Jean-Henri, H., Galgani, F., Collard, F., Goffart, A., Annual variation in neustonic micro- and meso-plastic particles and zooplankton in the Bay of Calvi (Mediterranean–Corsica), 2014.
18. Michael, H., Leda, P., Gary, L., Harry, S., *Computer modeling of pesticide transport in soil for five instrumented watersheds*, U.S. Environmental Protection Agency, Southeast Water laboratory, Athens, Ga. by ESL Inc., Sunnyvale, California, 1973.
19. Dara, S.S., *A text book of environmental chemistry and pollution control*, p. 219, S.Chand and company Ltd, INDIA, 2008.
20. U.S. Environmental Protection Agency, Risk Assessment Guidance for Superfund, in: *Human Health Evaluation Manual*, Office of Emergency and Remedial Response, Washington D.C. 20450, 1989.

21. George, R., Joy, V., Aiswarya, S., Jacob, P.A., Treatment Methods for Contaminated Soils, Translating Science into Practice. *Int. J. Educ. Appl. Res.*, 2016.
22. Ezio, R., Fabian, B., Petros, G., Bernardino, C., Soil Pollution Prevention and Remediation. *J. Appl. Environ. Soil Sci.*, 1, https://doi.org/10.1155/2016/9415175, 2016, Article ID 9415175, 2016.
23. Skala, J., Vacha, R., Cupr, P., Which compounds contribute most to elevated soil pollution and the corresponding health risks in flood plains in the head water areas of the central European watershed. *Int. J. Environ. Res. Public Health*, 15, 1146, 2018.
24. Retamel-Salgado, J., Hirzel, J., Walter, I., Matus, I., Bioabsorption and bio-accumulation of cadmium in the straw and grain of maize (*Zea mays* L.) in growing soils contaminated with cadmium in different environment. *Int. J. Environ. Res. Public Health*, 14, 1399, 2017.
25. Yang, K., Zhang, T., Shao, Y., Tian, C., Cattle, S.R., Zhu, Y., Song, J., Fractionation, bio accessibility and risk assessment of heavy metals in the soil of an urban recreational area amended with composted sewage sludge. *Int. J. Environ. Res. Public Health*, 15, 613, 2018.
26. Drainage Manual: A Guide to Integrating Plant, Soil, and Water Relationships for Drainage of Irrigated Lands, Interior Dept., Bureau of Reclamation, University Press of the Pacific, 2005.
27. Blaylock, A.D., *Soil Salinity and Salt tolerance of Horticultural and Landscape Plants*, University of Wyoming, Larmie, 1994.
28. Nijland, H.J. and El Guindy, S., Crop yields, water table depth and soil salinity in the Nile Delta, Egypt, in: *Annual report*, International Institute for Land Reclamation and Improvement, IL Wageningen, The Netherlands, 1983.
29. Sharma., D.P., Singh, K.N., Rao, K.V.G.K., Crop Production and soil salinity: evaluation of field data from India. Paper published in *Proceedings of the Symposium on Land Drainage for Salinity Control in Arid and Semi-Arid Regions*, Cairo, Egypt, February, 25th to March 2nd, 1990, vol. 3, pp. 373–383, 1990.
30. Oosterbaan, R.J., Crop yields, soil salinity and water table depth in Pakistan, in: *Annual Report 1981*, vol. 2, pp. 50–5429–33, International Institute for Land Reclamation and Improvement, Wageningen, The Netherlands, reprinted in Indus 24, 1990.
31. Paul, D., Osmotic stress adaptations in rhizobacteria. *J. Basic Microbiol.*, 53, 201252, 1–10, 2012.
32. Gray, E.J. and Smith, D.L., Intracellular and extracellular PGPR: commonalities and distinctions in the plant-bacterium signaling processes. *Soil Biol. Biochem.*, 37, 395, 2005.
33. Dimkpa, C., Weinand, T., Ash, F., Plant-rhizobacteria interactions alleviate abiotic stress conditions. *Plant Cell Environ.*, 32, 1682, 2009.

34. Hayat, R., Ali, S., Amara, U., Khalid, R., Ahmed, I., Soil beneficial bacteria and their role in plant growth promotion: a review. *Ann. Microbiol.*, 60, 579, 2010.
35. Lutgtenberg, B. and Kamilova, F., Plant-growth-promoting rhizobacteria. *Annu. Rev. Microbiol.*, 63, 541–55, 2010.
36. Blaylock, A.D., *Soil salinity Salt tolerance and growth potential of horticultural and landscape plants*, Co-operative Extension Service, University of Wyoming, Department of Plant, Soil and Insect Sciences, College of Agriculture, Laramie, Wyoming, 994.
37. Netondo, G.W., Onyango, J.C., Beck, E., Sorghum and salinity: II. Gas exchange and chlorophyll fluorescence of sorghum under salt stress. *Crop Sci.*, 44, 806, 2004.
38. Wang, W., Vinocur, B., Altman, A., Plant responses to drought, salinity and extreme temperatures: towards genetic engineering for stress tolerance. *Planta*, 21, 1, 2003.
39. Nia, S.H., Zarea, M.J., Rejali, F., Varma, A., Yield and yield components of wheat as affected by salinity and inoculation with *Azospirillum* strains from saline or non-saline soil. *J. Saudi Soc. Agric. Sci.*, 11, 113, 2012.
40. Ramadoss, D., Lakkineni, V.K., Bose, P., Ali, S., Annapurna, K., Mitigation of salt stress in wheat seedlings by halotolerant bacteria isolated from saline habitats. *Springerplus*, 2, 6, 1, 2013.
41. Yao, L., Wu, Z., Zheng, Y., Kaleem, I., Li, C., Growth promotion and protection against salt Stress by *Pseudomonas on* cotton. *Eur. J. Soil Biol.*, 46, 49, 2013.
42. Netondo, G.W., Onyango, J.C., Beck, E., Sorghum and salinity: II. Gas exchange and chlorophyll fluorescence of sorghum under salt stress. *Crop Sci.*, 44, 806, 2004.
43. Ashraf, M., Some important physiological selection criteria for salt tolerance in Plants. *Flora*, 199, 361–376, 2004.
44. Munns, R. and James, R.A., Screening methods for salinity tolerance: a case study with tetraploid wheat. *Plant Soil*, 253, 201, 2003.
45. Tester, M. and Davenport, R., Na^+ tolerance and Na^+ transport in higher plants. *Ann. Bot.*, 91, 503–507, 2003.

7

Fertilization and Fertilizer Types

İdris Karagöz

Department of Polymer Materials Engineering, Yalova University, Yalova, Turkey

Abstract

Nowadays, fertilizers are among the most important production factors of sustainable and modern agriculture. With the use of science, technical methods, and fertilizers with different contents in agriculture, the quality and quantity of agricultural products have increased significantly. In this study, it is aimed to compile information about fertilizers and to create a reference source for those working in the agricultural sector and those interested in the subject by using the literature information. In the study, the purpose of fertilization and its application methods were mentioned, and fertilizers were classified under two main headings: fertilizers containing organic matters and chemical fertilizers. In the use of fertilizers, conducting soil analysis, knowing the nutrients needed by the plant to be planted, the chemical composition of the fertilizer, climatic conditions such as temperature and precipitation, procedures previously applied to the soil, and conscious agriculture have been determined to be essential factors. It has been observed that plants planted in areas poor in terms of nutrients do not only affect plant development but also affect human and animal health with the consumption of these plants. In the future, it is predicted that chemical fertilizers will be used more than other types of fertilizers in agricultural production in order to grow more quality products in agriculture for the purpose of meeting the food need emerging in parallel to the increasing world population.

Keywords: Fertilizers, fertilization, agricultural, plant, sustainability, nutrient, chemical

Email: idris.karagoz@yalova.edu.tr

Inamuddin, Mohd Imran Ahamed, Rajender Boddula and Tariq Altalhi (eds.) Applied Soil Chemistry, (123–148) © 2021 Scrivener Publishing LLC

7.1 Introduction

The ability of plants to make organic matters from inorganic matters has enabled plants and plant products to be the basic nutrients of humans and animals from past to present [1]. Scientific studies on the healthy nutrition of plants and increasing product efficiency have been conducted for long years. It has been proven by scientific studies that humans and animals who have to be fed with unhealthy and undernourished plants and plant products obtained from them suffer from different diseases due to vitamin and mineral deficiencies [1–3].

In order to be healthy, humans and animals need 7 macro mineral elements (high requirement), such as calcium (Ca), phosphorus (P), potassium (K), sodium (Na), sulfur (S), magnesium (Mg), and chlorine (Cl), and 18 micro mineral elements (low requirement), such as cobalt (Co), chromium (Cr), iodine (I), molybdenum (Mo), zinc (Zn), copper (Cu), iron (Fe), manganese (Mn), selenium (Se), aluminum (Al), boron (B), fluorine (F), lead (Pb), nickel (Ni), silicon (Si), vanadium (V), arsenic (As), and bromine (Br) [1, 5]. Nowadays, zinc (Zn) and iron (Fe) deficiencies are observed most frequently in humans [4, 6]. Of Zn in the human body, 80% is in the skeleton and muscles around the skeleton. Zinc deficiency delays the healing of wounds in humans and animals [7]. According to the data of the World Health Organization (WHO), it is stated that iron deficiency is detected in approximately 3.7 billion people worldwide and iron deficiency anemia is detected in 2 billion people. Feeding with plants and plant products is shown as the main reason for this. Similarly, feeding animals with meadow-pasture plants that do not contain phosphorus turns bones into a soft and sponge-like porous structure and causes them not to develop normally. This situation, which leads to a desire to eat soil in animals, reduces milk yield by 30%, while it also causes problems, such as the loss of appetite, non-conception, and frequent miscarriage of offspring [1]. Nowadays, phosphorus deficiency in humans and animals is regarded as one of the most critical problems in many countries of the world since it affects development directly.

Plants take nutrients from the soil. Due to the intake of nutrients by plants from the soil, the removal of nutrients from the soil due to improper irrigation, the destruction of nutrients by erosion, and the evaporation of nutrients, soils become poorer in terms of plant nutrients over time [8]. It is not possible to take these elements from the soil continuously without giving anything to the soil. The agriculture performed in this way is called robber agriculture, and the amount of products obtained from the unit agricultural area decreases. In the soil impoverished in terms of plant nutrients, the product yield and amount are increased by enriching the soil in terms of nutrients by the use of fertilizers [9, 10]. In order for fertilization to be performed correctly, conducting soil analysis, knowing the nutrients needed by the plant to be planted, the effect of the climate, and conscious agriculture are essential. It is essential to select the right fertilizer and apply it to the soil in the right amount and at the right time in order to obtain maximum benefit from fertilizers without damaging the environment and to obtain quality and efficient products. The consumption of fertilizers has been regarded as an indicator of the agricultural development and wealth of countries for many years [1, 11].

In this study, it was aimed to contribute to the correct and appropriate use of fertilizers in agriculture, classification of fertilizer types, and increasing the product yield without damaging the environment by selecting a correct and appropriate fertilizer for the plant needs as a result of the literature review.

7.2 The Purpose and Application Methods of Fertilization

Nowadays, fertilizers are one of the most important input materials of agricultural production. There are many definitions in the literature regarding the terms of fertilizer and fertilization. However, any substance added to the soil in order to increase the yielding power of the cultural soils and increase the quality and quantity of the product is commonly defined as a fertilizer, and performing this procedure is defined as fertilization.

The primary purpose of fertilization is to obtain products of high quality and in high amount from the fertilizer applied to the unit agricultural area and generate income for the producer. However, while achieving this purpose, fertilizers should be applied in the amount and content needed by the plant and at the required time, by using the right method, protecting natural resources, and not damaging the environment, human, and animal health [1, 5, 12]. Plant, climate, and soil have a very significant effect on balanced and economical fertilization using a suitable method [8, 10, 12].

Plants take different amounts of nutrients in the environments where they develop depending on their genetic characteristics. However, the intake of nutrients is affected by the compatibility of the plant with its environment, its age, and the amount of CO_2 released by its root systems [13]. Cultivated plants planted for seeds or grains receive much fewer nutrients from the soil than cultivated plants harvested with their above-ground organs. The amount of nutrients the plant receives from the soil during its development periods increases. During these periods, the soil should be supported by the nutrients the plant needs. The root structure and root depth of the plant differentiate the nutritional needs of the plant. The plant with a superficial root system needs more nutrients than deep-rooted plants. The amount of the nutrient intake by the plant affects the development of the above-ground organs of the plant and the plant growth. At the beginning of the development period of cultivated plants, knowing the rooting pattern is an important factor in selecting the fertilizer application method and obtaining the maximum yield from the application [1]. Fertilizers should be applied at the beginning of the plant development, just below the seed or seed row if there is strong taproot development, and next to the seed row if there is a lot of lateral root development. Fertilizing the topsoil in an appropriate amount and cultivating it well enable the plant roots to grow in depth. The application of a fertilizer to the surface is shown schematically in Figure 7.1. Surface broadcast is a method applied across the entire area of the topsoil. In addition to its advantages such as a fast application by high-capacity fertilizer spreaders and being economical, it also has some disadvantages such as high nutrient loss and low homogeneity [14, 15]. Plants take the nutrients they need from regions where roots actively develop in the soil. Therefore, plants get the maximum benefit

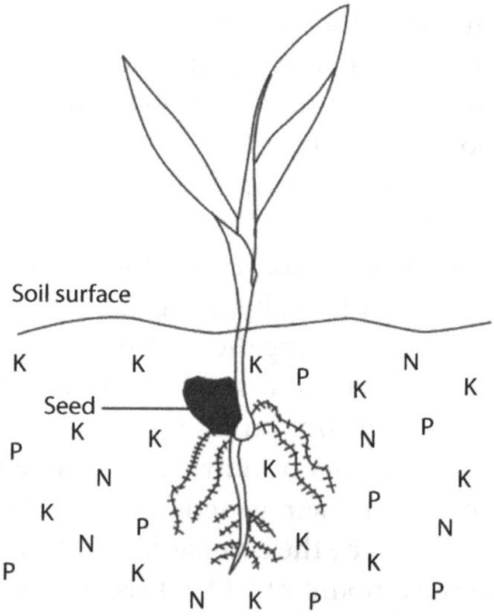

Figure 7.1 Surface broadcast application method.

from fertilization applied at the points close to roots, and they exhibit a better development.

The application of surface broadcast after planting is called topdress application. Its application on the soil surface before planting is called base dressing. The application of base dressing is easy and requires less labor. However, in regions with abundant fall precipitation, fertilizers should be applied as topdress to minimize loss by washing. The application of fertilizer for the maize plant on the soil surface with different methods is shown schematically in Figure 7.2. Each method has advantages and disadvantages compared to each other, which should be considered in selecting the method.

The band application is another method used for fertilization. In this method, fertilizers are applied to the bands that developing roots can easily access. As is seen in Figure 7.3, the fertilizer is commonly applied to a 5-cm side and depth of the seed or plant so that it would not be affected by salt or ammonium toxicity [1, 16].

The band application can increase nutrient use efficiency by providing plants with a concentrated nutrient area. The procedure can be performed before planting or seed application or simultaneously. Band application methods in the maize plant are shown schematically in Figure 7.4.

Temperature and precipitation are the main climatic factors. The temperature of the soil directly affects the root development of plants and the amount of nutrient utilization [16]. The intake of nutrients increases

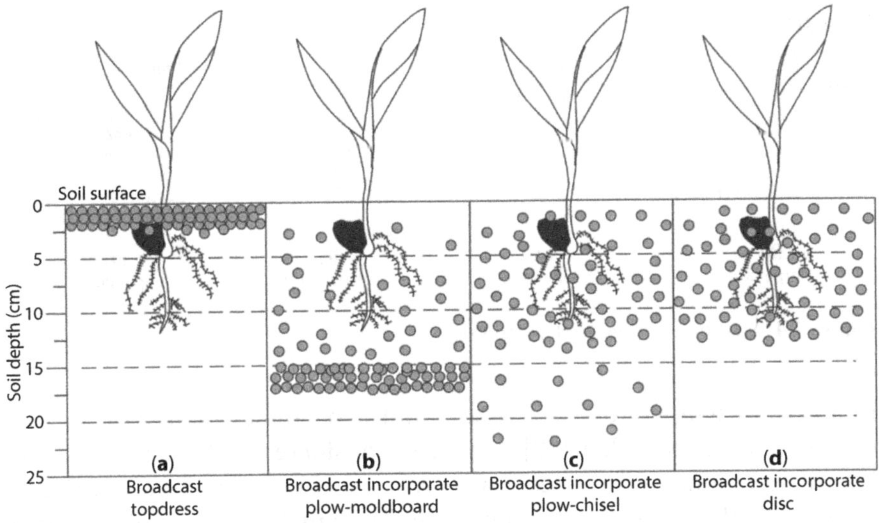

Figure 7.2 Methods used in mixing the fertilizers applied to the soil surface with soil.

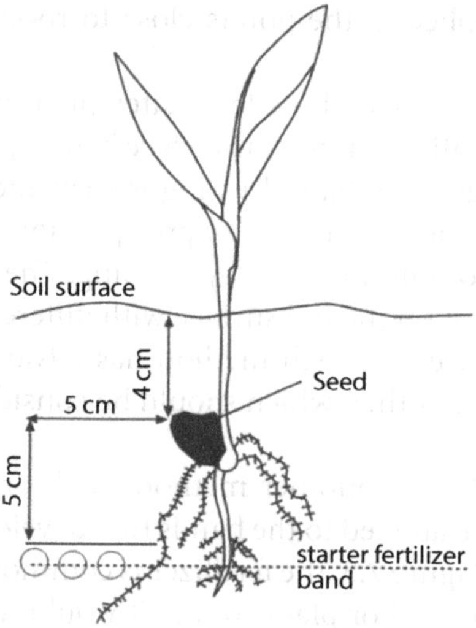

Figure 7.3 Band application method.

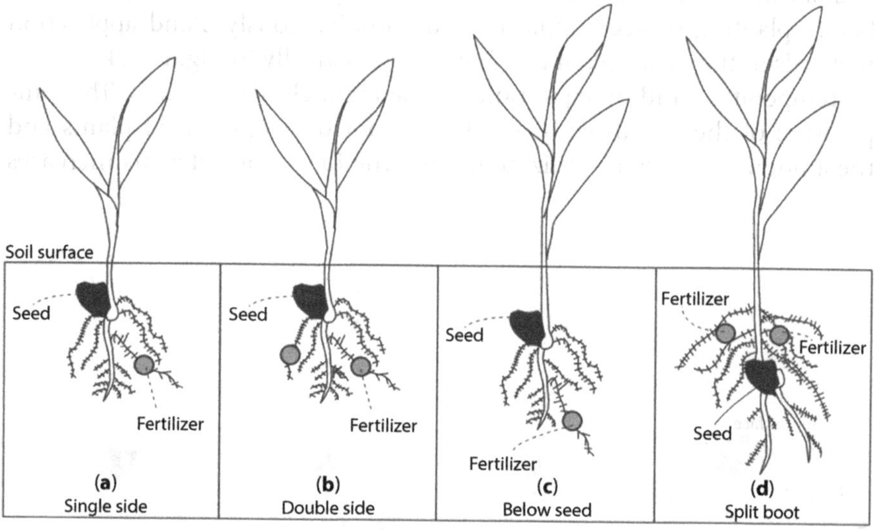

Figure 7.4 Methods of banding fertilizers.

to a certain temperature in the soil, and it decreases rapidly when this temperature is exceeded [17], which is considered to be caused by the loss of functions of enzymes that play a role in the intake of nutrients, along with the increasing temperature. The nutrient intake of the plant varies with the changing soil temperature according to seasons. While the subsoil warms later than the surface in spring, the exact opposite is the case in fall. Slow plant growth, especially in spring, results from the lesser intake of nutrients from the cold soil. The development of plant roots is better in dry soils compared to wet soils. It is quite important to know the efficiency of fertilizers in regions receiving precipitation. For example, it is known that phosphorus fertilizers increase root development in the plant and ripen plants quickly. The application of phosphorus fertilizers, especially in areas with precipitation, will increase the amount of product obtained by ripening the plant more quickly. Another critical point is the area to which fertilizers will be applied in the soil. The application of fertilizers to a suitable place according to seeds or plants is as important as applying the right amount of fertilizers. In arid and semi-arid regions, the application of fertilization should be performed at a depth where the water is excessive. In humid regions, fertilizers should be applied to a deep or deeper level depending on the type of nutrients in a way to provide maximum benefit for the plant.

To know the physical, chemical, and biological properties of the soil is essential for proper and economical fertilizer utilization [16, 17]. Plant roots need oxygen so that plants take nutrients adequately. The utilization ratio of nutrients is very low in soils that cannot be adequately aerated and that cannot be aerated for reasons such as waterlogging. It causes the poisoning effect on the root due to oxygen, which is not enough in the root zone. The pH of the soil is also effective on root growth [18]. The ratio of benefiting from fertilizers of the plant with the insufficient root growth also decreases. Two different elements in the soil may also affect each other's intake negatively. Similarly, two different elements may affect each other's intake positively. There is a close relationship between the chemical structure of the soil and fertilization to be applied [13]. The form of application and the procedures previously applied to the soil, etc., lead to the differentiation of yield. The procedures previously applied to the soil are, therefore, of great importance for balanced and economical fertilization.

Fertilizers are applied to the soil with different techniques for reasons such as 1- ensuring that the plant takes the nutrients it needs easily, 2- minimizing fertilizer loss and environmental damage, and 3- providing convenience in the application. The application of fertilization with an accurate technique will prevent the contamination of surface and groundwater. The following factors are influential in selecting the method in the application of fertilization [1].

1. The chemical composition of the fertilizer to be applied
2. Soil type, pH level

3. Features of the plant to be planted
4. Nutrients needed by the plant to be planted
5. Climatic conditions

Chemical fertilizers are applied in two different ways, to the soil or to the plant, by considering the factors mentioned above. The forms of application of chemical fertilizers to the soil or plant are shown graphically in Figure 7.5.

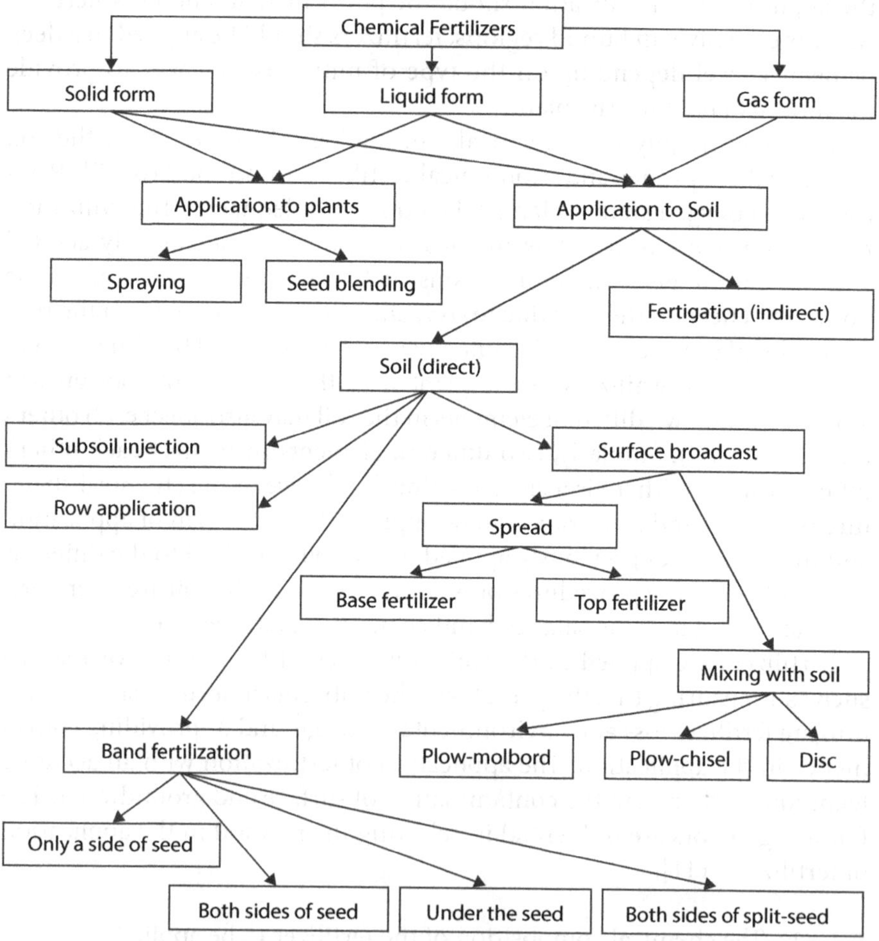

Figure 7.5 Chemical fertilizer application methods.

7.3 Classification of Fertilizers

With respect to the classification of fertilizers, in the literature, there are classification methods with the same basic characteristics but different words. Similarly to the classification made by the European Union (EU), fertilizers are commonly classified under two main groups: 1- fertilizers containing organic matters and 2- chemical fertilizers [1, 17, 19]. The classification of fertilizers are presented in Table 7.1.

Table 7.1 Classification of fertilizers.

Fertilizers	
1. Fertilizers containing organic matters	**2. Chemical fertilizers**
1.1. Herbal and animal-based fertilizers -Solid and liquid organic fertilizer -Solid and liquid farm yard manure -Poultry solid and liquid animal manure -Bat manure -Microbial **1.2. Organominerel fertilizers** -Organomineral fertilizer with N -Organomineral fertilizer with NP -Organomineral fertilizer with NK -Organomineral fertilizer with NPK -Enzyme additives -Other fertilizers **1.3. Soil conditioners** -Organic soil conditioners -Mineral soil conditioners -Organic + mineral soil conditioners	**2.1. Solid and liquid inorganic fertilizers with primary single and compound plant nutrients** 2.1.1. Straight fertilizers -Nitrogen fertilizers -Phosphate fertilizers -Potassium (potash) fertilizers 2.1.2. Compound or multinutrient fertilizers -NPK fertilizers -NP fertilizers -NK fertilizers -PK fertilizers 2.1.3. Inorganic liquid fertilizers -Liquid fertilizers with single plant nutrients -Liquid fertilizers with compound plant nutrients **2.2. Inorganic fertilizers with secondary nutrients** 2.2.1. Micronutrient fertilizers -Fertilizers containing boron (B) -Fertilizers containing cobalt (Co) -Fertilizers containing copper (Cu) -Fertilizers containing iron (Fe) -Fertilizers containing manganese (Mn) -Fertilizers containing molybdenum (Mo) -Fertilizers containing zinc (Zn)

Fertilizers that contain plant nutrients as organic compounds facilitate the intake of plant nutrients by improving the physical, chemical, and biological structure of the soil and are prepared from animal waste, plants or by-products obtained by the processing of plants are defined as fertilizers containing organic matters. Chemicals that are produced by chemical methods, contain plant nutrients, are found in granule, prill, pellet, crystal, powder, gas, solution or suspension, and are applied to leaves and above-ground parts by spraying to the soil through diluting or dissolving are defined as chemical fertilizers.

According to the EU standards, the word "EC FERTILIZER" must be written on fertilizer labels in capital letters and in a clearly visible way at the beginning of the label. The content of chemical fertilizers, type of fertilizer, and storage conditions should be written clearly and indelibly on labels. If a chemical fertilizer also contains microelements, the word "microelement" should be added to the type name [17].

Fertilizers are classified as solid and liquid fertilizers according to their physical properties. Solid fertilizers can be in the form of crystal, powder, prill, briquette, granules, and supergranules. Liquid fertilizers are applied through irrigation water or directly to the plant [16]. Their ease of use, low labor requirement, and easy mixing with herbicides affect farmers' preference for liquid fertilizers [20].

7.4 Fertilizers Containing Organic Matters

7.4.1 Herbal and Animal-Based Fertilizers

Soil contains organic matters by between 1% and 5%. However, organic matters that provide macro and micronutrients for the nutrition of plants are not considered as organic fertilizers alone [9]. Organic matters that have buffering properties against acidity, alkalinity, and salinity in the soil are effective on the biological, chemical, and physical properties of soils [19]. They are a source of carbon and energy for micro-organisms in the soil [21]. In order for a product to be an organic fertilizer, it should contain organic matters and plant nutrients to meet the requirement and microorganisms useful for the plant.

Solid organic fertilizers consisting of unprocessed herbal-based materials and partially animal-based materials (excluding fecal-based ones) are composed of high amounts of carbon (C) and plant nutrients. Liquid organic fertilizers are fertilizers in the form of solutions or suspensions obtained by the physical or chemical treatment of herbal-based materials and partially animal-based materials (excluding fecal-based ones) [1]. To know the C/N ratio of an organic fertilizer is important for the determination of the fertilizer value. The amount of organic matter should be minimum 40% in solid organic fertilizers and minimum 20% in liquid organic fertilizers. The total amount of $N+P_2O_5+K_2O$ should be minimum 3% for solid organic fertilizers and minimum 1% for liquid organic fertilizers. The amount of organic matter, total nitrogen, organic nitrogen, total (humic and fulvic) acid if more than 1%, total free amino acids if more than 2%, amount of P_2O_5, amount of water-soluble K_2O, C/N ratio and maximum humidity data should be available on the labels of solid and liquid organic fertilizers.

Farmyard manure basically consists of solid and liquid stools of bovine and small cattle in the farm and bedding laid on the floor [1, 22]. Solid farmyard manure is obtained by composting feces and removing moisture, and liquid farmyard manure is obtained by dissolving solid farmyard manure in water. Unlike chemical fertilizers, farmyard manure provides the soil with the nutrients required for the plant and also ensures that the soil structure is suitable for agriculture. Farmyard manure increases the water-holding capacity of the soil and ensures the aeration of the soil by facilitating the cultivation of the soil [22]. Most of the plant nutrients in its content are water-soluble. Its effect usually lasts for 3–4 years. The time, form, and amount of application vary depending on the structure of the soil, plant type, environmental conditions, and the application method, time, and amount of chemical fertilizers.

Fertilizers obtained by the aerobic composting of poultry manure with or without base or by maturing the feces of other winged animals, such as bats, in their natural environment or aerobic composting of them, and reducing the moisture are named as solid poultry manure. Liquid fertilizers obtained by dissolving poultry manure in water are named as liquid poultry manure [1]. Chicken manure usually first comes to mind when poultry manure is mentioned. Chicken manure is richer in terms of organic matter and plant nutrients since solid and liquid feces exist together compared to farmyard manure [23]. Since its smell is generally unpleasant for the environment, it should be stacked without waiting and composted by a suitable method. By composting, it gains a feature applicable to all types of soil and all plants. Bat manure was commonly used in agriculture in the USA toward the end of the 1800s. Bat manure is rich in plant nutrients and contains a large amount of micro-organisms and humic and fulvic acid by 20%–30%. The microorganisms in its content mineralize the organic compounds in the soil by easily breaking them down. It is essential to prevent the damage caused by fulvic acid, nematodes in the plant and the diseases caused by fungi in the plant [1].

Commercially formulated products of living microorganisms that play a role in providing the nutrients necessary for the plant and making them biologically useful are called microbial fertilizers [24]. Microbial fertilizers improve the quality and quantity of the product by increasing the amount of nutrients useful for the plant [25]. They do not pose a risk for human and animal health. They are applied by being grafted into the seed. They are grafted into the seed by dry grafting, wetting the seed, and using an adhesive (metal cellulose, milk powder, gum arabic, etc.).

7.4.2 Organomineral Fertilizers

Solid and liquid fertilizers obtained by the reaction or mixture of one or more organic fertilizers with one or more primary single, compound, or micronutrient chemical fertilizers are defined as organomineral fertilizers. In other words, organomineral fertilizers are obtained by a chemical reaction to organic fertilizers or by mixing them directly with chemical fertilizers [26]. Thus, the efficiency of a fertilizer in traditional agriculture is increased by making the fertilizer content rich in plant nutrients. Similarly, the addition of chemical fertilizers to organic fertilizers also prevents the loss of plant nutrients. Nowadays, preservatives are added to organic fertilizers mainly for preventing nitrogen losses in animal feces [1]. Phosphoric, sulfur, and hydrochloric acids prevent the breaking down of urea by making the reaction of fertilizer acid. Therefore, these acids are effective preservatives. The fact that they are expensive and their limited use are the disadvantages of these acids. Superphosphate containing calcium salts such as $CaSO_4$, $CaCl_2$, $CA(NO_3)_2$, and P_2O_5 by 16%–18% are also used as a preservative.

While the organic matter content of solid organomineral fertilizers with N should be minimum 20%, their nitrogen content should be minimum 12% and maximum moisture content should be 20%. While the organic matter content of solid organomineral fertilizers with NP should be minimum 20%, their nitrogen content should be minimum 3%, P_2O_5 should be minimum 5%, $(N+P_2O_5)$ should be minimum 12%, and the maximum moisture content should be 20% [17]. While the organic matter content of solid organomineral fertilizers with NK should be minimum 20%, their nitrogen content should be minimum 3%, water-soluble K_2O should be minimum 5%, $(N+K_2O)$ should be minimum 12%, and the maximum moisture content should be 20%. While the organic matter content of solid organomineral fertilizers with NPK should be minimum 20%, their nitrogen content should be minimum 3%, water-soluble K_2O should be minimum 5%, ($N+P_2O_5+K_2O$) should be minimum 15%, the and maximum moisture content should be 20%.

While the organic matter content of liquid organomineral fertilizers with N should be minimum 15%, their nitrogen content should be minimum 8%. While the organic matter content of liquid organomineral fertilizers with NP should be minimum 15%, their nitrogen content should be minimum 3%, P_2O_5 should be minimum 4%, ($N+P_2O_5$) should be minimum 10%. While the organic matter content of liquid organomineral fertilizers with NK should be minimum 15%, their nitrogen content should be minimum 3%, water-soluble K_2O should be minimum 4%, and ($N+K_2O$) should be minimum 10% [17]. While the organic matter content of solid organomineral fertilizers with NPK should be minimum 15%, their nitrogen content should be minimum 2%, P_2O_5 should be minimum 4%, water-soluble K_2O should be minimum 4%, and ($N+P_2O_5+K_2O$) should be minimum 12%. Enzyme-added products are defined as the products containing biocatalysts that break down the proteins, carbohydrates, and fats within the plant to their smallest units [1]. Enzyme-added products break down the proteins, fats, starches, and celluloses in the plant cell walls and facilitate their intake by the plant. They can be of herbal, animal, and microbial origin. Other organomineral fertilizers are the products such as organic fertilizers obtained by the coating of inorganic fertilizers with NPK, vinas extract and liquid vinas, solid and liquid seaweed, liquid humic acid or liquid fulvic acid, and fermentation [1, 17].

7.4.3 Soil Conditioners

Soil conditioners are classified as organic soil conditioners, mineral soil conditioners, and organic added mineral soil conditioners. In the literature, they are observed to be used as soil conditioners in some polymer materials [27]. When soil conditioners are used properly, they ensure the improvement of soils that are not suitable for agriculture and make them suitable for agriculture by improving the properties of the soil [28, 29].

Organic soil conditioners are divided into three groups: plant compost, domestic compost, and leonardite. While plant compost is obtained by the fermentation of plant organic wastes, domestic compost is obtained by the fermentation of domestic organic wastes. Both of them should contain minimum 35% organic matter and maximum 20% moisture content, and 90% of the product should be sieved through a 10-mm sieve. Leonardite, which is a natural organic matter consisting of macro and micronutrients as well as a high content of humic acids and carbon, is an organic clay sedimentary rock stratified as a result of the humidification and degradation of sedimentary plant and animal residues from ancient times in lakes and marsh environments due to pressure and temperature [29]. It can be used as liquid and solid in agriculture. In the content of soil conditioners made of leonardite, the humic and fulvic acid amount should be minimum 40%, and $CaCO_3$ should be maximum 5%. The granule size should be between 1 and 4 mm, and the powdered grain size should be maximum 50 mesh.

Mineral soil conditioners such as unslaked and slaked agricultural lime, gypsum, dolomite, zeolite group minerals, vermiculite, diatomite, and pumice stone are used to create a suitable environment for plant growth and to balance the pH of the soil [1, 5]. While agricultural lime is used to bring the acidity of the soil to a certain level, gypsum is used for the improvement of sodium in alkaline soils, dolomite, diatomite, and zeolite minerals are used to increase the water-holding capacity of the soil and to make the soil moist, airy, and permeable, vermiculite is used to accelerate plant root growth and development, and perlite is used for soilless agriculture applications and water-saving [1, 29].

The use of organic added mineral soil conditioners is limited. They are prepared as a mixture in the event that the desired results are not achieved by using organic or mineral soil conditioners alone. In general, they are mixed into the soil in appropriate proportions, and they ensure the transformation of the soil into a suitable environment for plant growth.

7.5 Chemical Fertilizers

The first patent for the production of chemical fertilizers was obtained in 1849. Primary and secondary nutrients are mentioned in the patent content, similar to common classifications nowadays [30]. Chemical fertilizers can be defined as any inorganic material which is wholly or partially of synthetic origin and added to the soil to support plant growth. Chemical fertilizers can be found in solid and liquid forms. However, they are generally preferred as granules for the ease of transportation and storage. They are intensely used in agricultural applications compared to other types of fertilizers [10]. Chemical fertilizers are rich in three essential nutrients (nitrogen, phosphorus, and potassium) that are needed by plants and contain one or more plant nutrients in their composition [5]. Unlike organic fertilizers, they contain high amounts of plant nutrients and are easily soluble in water. It is very important to know the characteristics of chemical fertilizers so that they can be used effectively, economically, and consciously. The amount of agricultural production and the quality of the product increase along with their correct and effective use. Commercially used chemical fertilizers are reliable since their content is known. Different mixtures can be prepared according to the nutrients needed by plant species. However, unconscious use affects the properties of the soil and leads to changes in salinity and acidity content in the soil and the pollution of surface and groundwater [31]. The soil health is deteriorated and production costs increase as a result of the intensive use of them by farmers to obtain more products. Some types have high acid content or toxic substances that may burn human skin and adversely affect the respiratory system. Therefore, the proper use of chemical fertilizers consciously and according to the needs of the plant is quite crucial for sustainable agriculture.

Fertilizers that contain one or more plant nutrients in their compositions and are produced by chemical methods are defined and classified as chemical fertilizers. Chemical fertilizers are classified into three main groups: solid and liquid fertilizers with inorganic primary single and compound plant nutrients, fertilizers with inorganic secondary nutrients, and fertilizers with micronutrients [1]. According to the EU standards, the word "EC FERTILIZER," fertilizer type name, the name of the nutrient contained, chemical symbol, and the amount in percent should be provided on the labels of chemical fertilizers. If there are microelements in the content of chemical fertilizers, the word "Microelements" should be added to the type name, and they should be listed alphabetically according to the chemical symbols.

The plant nutrients contained in chemical fertilizers are classified in three groups: 1- primary plant nutrients (nitrogen, phosphorus, and potassium), 2- secondary plant nutrients (calcium, magnesium, sodium, and sulfur), and 3- micro plant nutrients (boron, copper, cobalt, iron, manganese, molybdenum, and zinc). Chemical fertilizers are classified into three main groups according to the plant nutrients they contain [1, 5]. They are as follows:

1. Solid and liquid fertilizers with inorganic primary single and compound plant nutrients
2. Fertilizers with inorganic secondary nutrients
3. Fertilizers with micro plant nutrients

7.5.1 Solid and Liquid Fertilizers With Inorganic Primary Single and Compound Plant Nutrients

Nitrogen, phosphorus, and potassium are the three main nutrients needed by plants [30]. The sufficient amount of water and food sources directly affect the product yield. Nitrogen has a significant effect on product efficiency, and the growth rate of the plant directly varies by the ratio of N received by the plant [8]. More than one raw materials such as natural gas, naphtha, pyrite, coal, sulfuric acid, ammonia, and nitric acid are used in the production of nitrogen fertilizers [5]. Nitrogen chemical fertilizers usually contain nitrogen in the form of nitrate (NO_3) and ammonium (NH_4). It is important that the soil is aerated and drained in the plant's nitrogen intake from the soil. Plants take nitrogen in the form of NO_3- in well-aerated soils and in the form of NH_4+ in the soils lacking oxygen. The pH level of the soil, soil water retention power, and ambient temperature also have great importance on nitrogen intake. Nitrogen fertilizers should not be applied with seed in places where soil moisture is low [32]. Most plants need more nitrogen during vegetative growth periods compared to the growth period. In general terms, the factors that increase the nitrogen use efficiency of plants can be listed as follows [1]:

1. Calcification of acid soils
2. Crop rotation
3. Green fertilization
4. Farmyard manure application
5. Providing adequate water
6. Application of nitrogen fertilizers suitable for the plant at the appropriate time, in a sufficient amount, and with a suitable method
7. Use of nitrification inhibitors
8. Fighting weeds, diseases, and pests

Ammonium nitrate (AN) fertilizer has an off-white and granular structure with high water solubility. It is a product preferred by fertilizer producers [5]. It is produced by the neutralization of nitric acid with ammonia gas [33]. Half of the nitrogen in the fertilizer is nitrate (NO3) and half is ammonium (NH4). AN fertilizers with a nitrogen content of more than 28% N have the potential to burst under pressure. Therefore, fertilizer bags should not be stored more than four bags in a row while storing them. The following issues should be considered in the production of fertilizers with high nitrogen content:

1. The oil retention ratio of fertilizers should not exceed 4%.
2. The amount of carbon should not exceed 0.2% for fertilizers containing a minimum of 31.5% N and 0.4% for fertilizers containing 28%–31.5% N.

3. The pH value of 10 g fertilizer solution that is soluble in 100-ml water should be minimum 4.5.
4. More than 5% of it should not pass through a 1-mm sieve, and more than 3% should not pass through a 0.5-mm sieve.
5. It should contain maximum 0.01% Cl.
6. The amount of copper should not exceed 10 mg/kg.
7. In the explosion-proof test, the compression of at least one lead cylinder should not exceed 5% in each ignition process.

Calcium (lime) AN (CAN) is obtained by the neutralization of nitric acid with anhydrous ammonia in order to eliminate the possibility of an explosion in AN fertilizers [5]. Thirty-five to forty percent dolomite, calcium carbonate, and ground limestone are also mixed into the solution. Approximately, 16% of fertilizers produced all over the world are composed of AN and CAN. Production of solid nitrogen fertilizers are shown schematically in Figure 7.6.

Ammonium sulfate (AS) is produced by the reaction of ammonia gas and sulfuric acid and is also usually known as sugar fertilizer due to its

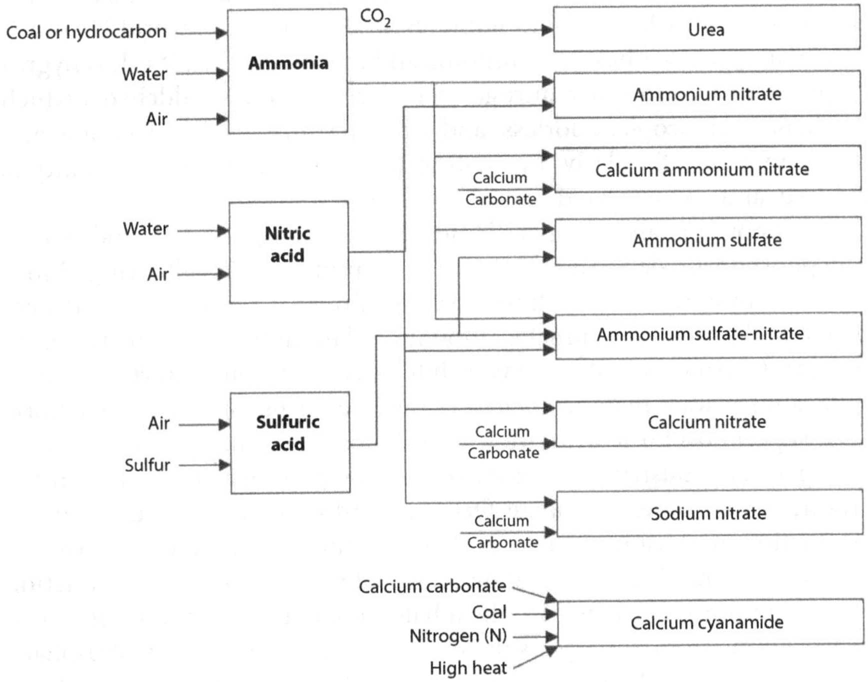

Figure 7.6 Production of solid nitrogen fertilizers.

white color and crystal structure. It is less soluble in water compared to urea. The N content should be minimum 19.7%. AS-nitrate fertilizer, which is the double salt of AS and AN, usually has a white and granular structure. It should contain a minimum of 25% N according to the principle of weight. It contains minimum 5% nitrate nitrogen and 5% sulfur (S). AN contains 60% more nitrogen compared to AS [33].

Calcium nitrate is mostly produced in the EU countries [1, 5]. It is obtained by the treatment of limestone with nitric acid. It should contain minimum 15% N, 21% Ca, and maximum 1.5% ammonia nitrogen. Sodium nitrate, which is a white and odorless fertilizer, is produced through chemical means. The fertilizer should contain minimum 15% N and 26% Na. Calcium cyanamide, which is bluish-black or livid, is obtained by the conversion of nitrogen in cyanamide into ammonia gas and the conversion of it into nitric acid. It is intensely used in Germany [31]. Calcium cyanamide is degraded in the soil by hydrolyzing with water, and urea is formed. The intermediate products formed by urea harm the plant by the poisoning effect. Therefore, calcium cyanamide should be mixed with the soil at least 2–3 weeks before planting, and it should be ensured that intermediate products are transformed into products before planting.

Urea is a white-colored and odorless fertilizer with high levels of nitrogen (min 44%), which is commonly used all around the world [10]. It is obtained by the synthesis of ammonia and carbon dioxide. It is less hygroscopic compared to other nitrogen fertilizers. Urea-formaldehyde, which is a light hygroscopic, odorless, and white fertilizer, is dissolved less and slowly in water. Similarly, by coating urea with sulfur, the solubility is reduced, and sulfur-coated urea is obtained.

Phosphorus is an essential element for humans, animals, and plants, and phosphorus deficiency causes major problems for all living things [5]. Therefore, the use of phosphorus fertilizers is compulsory and necessary. Soil properties, climate, amount of plant nutrients in the fertilizer, and plants grown should be taken into account in phosphorus fertilizer applications. Raw materials such as raw (rock) phosphates, phosphoric acid, superphosphoric acid are used in phosphorus fertilizers [5, 30]. In the fertilizer industry, while 60% of the raw phosphates are used in the production of superphosphate fertilizers, 10% of them are used in the production of phosphoric acid, and 10% of them are used in the production of compound fertilizers. Raw phosphates are used in the production of ammonium phosphate, AS phosphate, mono and diammonium phosphate, ammonium superphosphate, magnesium ammonium phosphate, calcium metaphosphate, and calcined phosphate-like fertilizers. The basic philosophy in the production of phosphorous solid, liquid and compound fertilizers is the conversion of phosphorus into a useful form for the plant. The factors that increase the efficiency of phosphorus use in plants can be listed as follows [1]:

1. Chemical compounds in the rhizosphere increase in quantity
2. The plant root system develops in accordance with the conditions
3. Useful micro-organism activity increases in the root domain
4. The genetic structure of the plant is suitable for increasing phosphorus use efficiency

Potassium fertilizers have been used since 1861 [30]. Raw materials used for the production of potassium fertilizers should have high soluble potassium content. Potassium fertilizers are the primary source of potassium of plants. Herbal and animal-based wastes and farmyard manure are good sources of potassium. The main differences that distinguish potassium fertilizers from each other are the anions they have. The nutrient contents of commercially produced potassium chemical fertilizers are presented in Table 7.2. Potassium chloride and potassium sulfate are commonly used in agriculture due to their low price. Furthermore, potassium nitrate and potassium magnesium sulfate fertilizers are also commercially produced. However, the application of potassium nitrate is limited since it is expensive. In agriculture, potassium fertilizers are used to overcome sulfur, chlorine, and magnesium deficiencies, as well as potassium deficiency.

Table 7.2 Commonly used potassium fertilizers.

	Nutrients (%)					
Potassium fertilizer type	**K_2O**	**N**	**P_2O_5**	**S**	**Mg**	**Cl**
Potassium chloride	60–62	-	-	-	-	45-48
Potassium sulfate	50–52	-	-	17	-	-
Potassium nitrate	44	13	-	-	-	-
Potassium magnesium sulfate	22	-	-	22	11	-
Potassium sodium nitrate	14	15	-	-	-	-
Potassium polyphosphate	22–48	-	40–60	-	-	-
Potassium metaphosphate	40	-	55–60	-	-	-

Potassium chloride, which is easily soluble in water and contains 60%–62% K_2C and 45%–48% Cl, is usually white and has a crystal structure. Potassium chloride is produced from sylvinite (KCl.NaCl). Potassium sulfate, which is usually used as a source of potassium and sulfur, contains 50%–52% K_2C and 17% S. The chlorine content in potassium sulfate should be maximum 3%. Potassium sulfate fertilizer is produced from kainite ($MgSO_4.KCl.H_2O$), chenite ($K_2SO_4.MgSO_4.6H_2O$), leonite ($K_2SO_4.MgSO_4.4H_2O$), langbeinite ($K_2SO_4.2MgSO_4$), glaserite ($3K_2SO_4.Na_2SO_4$), and polyhalite ($K_2SO_4.MgSO_4.2CaSO_4.2H_2O$) minerals. Potassium nitrate fertilizer, also known as saltpeter, contains minimum 44% K_2O and 13% N. The chlorine content in potassium nitrate should be maximum 0.1%. Potassium nitrate can be produced as a result of the decomposition of KCl and NaNO3 mixture, it can also be produced as a result of the reaction of NHO_3 with KOH. Potassium-magnesium sulfate is mostly produced from langbeinite. Potassium-magnesium sulfate produced from langbeinite contains 22% K_2O, 18% MgO, and 22% S. The chlorine content in potassium-magnesium sulfate should be maximum 2.5%. It is also produced from kieserite in Europe and especially in Germany [5]. The use of potassium metaphosphate (KPO_3) containing 60.1% P_2O_5 and 39.9% K_2O as a fertilizer is limited for economic reasons. Quite good results were achieved in the trials performed with plants in fields and greenhouses. It is considered that it will be used more intensely along with the cheapening of production costs.

7.5.2 Fertilizers With Inorganic Secondary Nutrients

Calcium fertilizers, magnesium fertilizers, and sulfur fertilizers are classified as fertilizers with inorganic secondary nutrients [33]. The amount of secondary nutrients in plants is higher compared to primary nutrients (N, P, and K) and micro plant nutrients (B, Cu, Co, Fe, Mn, Mo, and Zn). Therefore, plants do not need secondary nutrients as much as primary plant nutrients.

Growth disorders are observed in plants with calcium deficiency [34]. In case of calcium deficiency in plants, root development and the amount of secretion in root tips decrease significantly, stem development is weak, and plants are more easily caught with fungal and parasitic diseases [35]. Calcium deficiency is attempted to be removed by liming, fertilizing, and mineral soil conditioners. In acid reaction soils, liming materials (limestone, dolomite limestone, quicklime, and blast furnace slag) are applied to reduce the acidity of the soil. Calcium fertilizers (calcium nitrate, CAN, calcium chloride, calcium cyanamide, raw phosphate, normal superphosphate, and triple superphosphate) are generally used in the form of nitrogen and phosphorus fertilizers. In the improvement of problematic sodic soils, gypsum ($CaSO_4.2H_2O$) is used as a mineral soil conditioner. Gypsum is also used as a fertilizer in low acid reaction soils since it contains calcium and sulfur.

Magnesium is a plant nutrient that plays a role in many enzyme activities and structural stabilization of tissues. Shorter roots, smaller size shoots, and necrotic spots appear in plants with magnesium deficiency [36]. Magnesium fertilizers are used to eliminate magnesium deficiency, which is generally observed in soils with high precipitation, high acid reaction, and coarse texture. Dolomite, dolomitic limestone, magnesium lime, magnesium sulfate, magnesium oxide, and potassium magnesium sulfate are commonly used in order to bring the pH of acid reaction soils to suitable proportions and to provide the Mg needed by the plant. If it is used in the form of irrigation or spraying, magnesium sulfate, magnesium chloride, magnesium nitrate, and synthetic clay should be preferred.

Although it is a macronutrient that all living organisms need, the studies on the sulfur need of plants are limited [37]. Sulfur has nitrogen-like effects on its uptake by plants, immobilization with microorganisms, and plant growth. Sulfur is present in the organic and inorganic forms in soils. The sulfur needed by the plant is provided from the atmosphere, other additives, fertilizers in a way to meet the needs of cultivated plants, and sulfur does not pose a problem for plants and the soil. However, sulfur deficiency that occurs in the soil due to various reasons constitutes a major problem for agriculture. Fertilizers containing elemental sulfur (ammonium phosphate-sulfur, ammonium-polyphosphate, ammonium phosphate urea-sulfur, and potassium chloride urea-sulfur), fertilizers containing sulfur in the form of sulfate (superphosphate, magnesium sulfate, manganese sulfate, potassium sulfate, etc.), and fertilizers containing sulfur in other ways (ammonium thiosulfate, ammonium bisulfate, ammonium polysulfate, and sulfur dioxide) can be shown as examples of sulfur fertilizers. The compounds containing sulfur in the form of sulfate are the primary source of sulfur as fertilizers.

7.5.3 Fertilizers With Micro Plant Nutrients

Ferrous fertilizers, manganese fertilizers, zinc fertilizers, copper fertilizers, boron fertilizers, and molybdenum fertilizers are classified as fertilizers with micronutrients [38, 39]. Iron deficiency constitutes a great problem since it is significant in the growth and development of plants [40]. According to the EU standards, water-soluble iron salt containing 12% Fe, iron chelate containing 5% Fe, and ferrous fertilizer solution containing 2% Fe are classified as ferrous fertilizer.

Manganese is an essential element for plant growth [41]. Manganese deficiency is more common in podzolic forest soils with poor drainage and coarse texture. The pH of the soil is one of the most critical factors in manganese intake [39]. According to the EU standards, there are five different manganese fertilizers, including manganese salt (17% Mn), manganese chelate (5% Mn), manganese oxide (40% Mn), and manganese-based fertilizer and fertilizer solution (3% Mn).

Although zinc is slightly needed by plants, it is very significant in plant development, photosynthesis, and sugar formation, starch, and sucrose synthesis. The intake of zinc by the plant is important in gene expression and regulation, root growth and development, and blooming and fruit formation. The photosynthesis performed by the plant decreases by between 50% and 70% depending on zinc deficiency, plant type, and the environment [1, 42]. According to the EU standards, there are five types of zinc fertilizer classification, including zinc salt (15% Zn), zinc chelate (5% Zn), zinc oxide (70% Zn), zinc-based fertilizer (30% Zn), and zinc fertilizer solution (3% Zn).

Among the micronutrients, copper is the least absorbed microelement by plants [43]. In case of copper deficiency, it is known that tuber (nodule) formation in legume plants, and formation of cellulose and hemicellulose, especially lignin, in the cell walls of higher plants decrease. Copper sulfate and copper ammonium phosphate fertilizers are commonly used to overcome copper deficiency since their water solubility is high, and they are cheap and easily available. According to the EU standards, there are eight different copper fertilizers, including copper salt (20% Cu), copper oxide (70% Cu), copper hydroxide (45% Cu), copper chelate (59 Cu), copper-based fertilizer (5% Cu), copper-based fertilizer solution (3% Cu), copper oxychloride (50% Cu), and copper oxychloride suspension (517 Cu).

In agricultural soils, boron deficiency is observed due to insufficient boron minerals in the soil and excessive calcification. Boron is effective in the root growth and growth of plants. Boron deficiency causes the pollen of plants to become poor, their seeds to become weak, wrinkled and deformed, and the number and size of flowers to decrease [44, 45]. According to the EU standards, there are six different boron fertilizers, including boric acid (14% B), sodium borate (10% B), calcium borate (7% B), boron ethanolamine (8% B), borate fertilizer in solution (2% B), and borate fertilizer in suspension (2% B).

The studies on molybdenum fertilizers are very few compared to Zn, Cu, B, Fe, and Mn fertilizers [1]. Unlike other fertilizers with micronutrients, molybdenum becomes more useful for plants as the pH value of the soil increases [39]. Molybdenum fertilizers are applied to the soil directly, to plants by spraying, and to the seed in a powder form. According to the EU standards, there are four different molybdenum fertilizers, including sodium molybdate (35% Mo), ammonium molybdate (50% Mo), molybdenum-based fertilizer (35% Mo), and molybdenum-based fertilizer solution (3% Mo).

7.6 Conclusion and Evaluation

While plants take carbon, hydrogen, and oxygen from air and water, they take all other nutrients from the soil. Plant nutrients in the soil can meet the needs of the plant to a certain extent. However, the nutrients in the soil decrease or run out depending on reasons such as permanent agricultural practices, rain, erosion, and wrong irrigation. It is essential to use fertilizers correctly and effectively in agricultural soils for sustainable agriculture and product efficiency. It is observed from the literature reviews that chemical fertilizers are used more intensely in agricultural applications compared to other types of fertilizers. Farmers should have knowledge about the plants they grow, soil and fertilizers to be used, and they should also be equipped to choose the fertilizer suitable for the plant so that fertilization can be performed correctly and effectively. The unconscious use of fertilizers adversely affects agricultural soils, environment, human and animal health, as well as product yield and quality.

References

1. Kaçar, B., *Temel Gübre Bilgisi*, Nobel Akademik Yayıncılık Eğitim Danışmanlık Tic. Ltd. Şti., Ankara, 2013.
2. Jha, P., Ram, M., Khan, M.A. *et al.*, Impact of organic manure and chemical fertilizers on artemisinin content and yield in *Artemisia annua* L. *Ind. Crops Prod.*, 33, 296–301, 2011.
3. Kaçar, B., *Bitki Besleme*, Ankara Üniversitesi Ziraat Fakültesi Yayınları No:899, Ankara Üniversitesi Basımevi, Ankara, 1984.
4. Welch, R.M. and Graham, R.D., Breeding for micronutrients in staple food crops from a human nutrition perspective. *J. Exp. Bot.*, 55, 353–364, 2004.
5. Pratt, C.J., Chemical fertilizers. *Sci. Am.*, 212, 6, 62–75, 1965.
6. Prasad, B. and Sinha, N.P., Changes in the status of micronutrients in soil with long term applications of chemical fertilizers, lime and manure. *Plant Soil*, 64, 437–441, 1982.

7. Prasad, M.N.V., *Trace Elements as Contaminants and Nutrients: Consequences in Ecosystems and Human Health*, John Wiley & Sons, Inc., Haboken, New Jersey, 2008.
8. Bøckman, O.C. and Olfs, H.W., Fertilizers, agronomy and N_2O. *Nutr. Cycling Agroecosyst.*, 52, 165–170, 1998.
9. Hendawy, S.F. and Khalid, K.A., Effect of chemical and organic fertilizers on yield and essential oil of chamomile flower heads. *Med. Aromat. Plant Sci. Biotechnol.*, 5, 1, 43–48, 2011.
10. Rai, N., Ashiya, P., Rathore, D.S., Comparative study of the effect of chemical fertilizers and organic fertilizers on *Eisenia foetida*. *IJIRSET*, 3, 5, 12991–12998, 2014.
11. Norton, G., Alwang, J., Masters, W., *Economics of Agricultural Development*, Routledge, London, 2015, https://doi.org/10.4324/9780203076217.
12. Eraslan, F., İnal, A., Güneş, A. *et al.*, Türkiye' de kimyasal gübre üretim ve tüketim durumu, sorunlar, çözüm önerileri ve yenilikler, TMMOB Zirat Müh. Odası, Ziraat Müh. VII. *Teknik Kongresi*, Ankara, 11–15 Ocak, 2010.
13. Nohrstedt, H.Ö., Arnebrant, K., Baath, E., Söderström, B., Changes in carbon content, respiration rate, ATP content, and microbial biomass in nitrogen-fertilized pine forest soils in Sweden. *Can. J. For. Res.*, 19, 323–328, 1989.
14. Cornell University, Describe advantages and limitations of the following fertilizer placement methods, https://nrcca.cals.cornell.edu/nutrient/CA4/CA0434.php, 2019.
15. Ombodi, A. and Saigusa, M., Broadcast application versus band application of polyefin-coated fertilizer on green peppers grown on adisol. *J. Plant Nutr.*, 23, 10, 1485–1493, 2000.
16. Mahler, R.L., *Fertilizer Placement CIS 757*, University of Idaho, University of Idaho Extension, https://www.extension.uidaho.edu/publishing/pdf/cis/cis0757.pdf, Moscow, 2019.
17. Wiley-VCH, (Ed.), *Ullmann's Agrochemicals, Fertilizers*, vol. 1, Wiley-VCH Verlag GmbH&Co. KGaA, Weinheim, 2007.
18. Harter, R.D., Effect of soil pH on adsorption of lead, copper, zinc, and nickel1. *Soil Sci. Soc. Am. J.*, 47, 47–51, 1983.
19. Tarım Kütüphanesi, Classification of fertilizers, http://www.tarimkutuphanesi.com, 2019.
20. Koli, P., Bhardwaj, N.R., Mahawer, S.K., *Agrochemicals: Harmful and benefical effects of climate changing scenarios*, pp. 65–94, Elsevier Inc., Amsterdam, Netherlands, 2019.
21. Powlson, D.S., Hirsch, P.R., Brookes, P.C., The role of soil microorganisms in soil organic matter conservation in the tropics. *Nutr. Cycling Agroecosyst.*, 61, 45–51, 2001.
22. Demirtaş, İ., Arı, N., Arpacıoğlu, A., Kaya, H., Özkan, C., Chemical properties of different organic manures. *Derim*, 22, 2, 47–52, 2005.

23. İnal, A., Sözüdoğru, S., Erden, D., Composition of poultry manure and its manurial value. *J. Agric. Sci.*, 2, 3, 45–50, 1996.
24. Zhang, L.Y., Recombinant microbial fertilizer and methods for its production, US Patent 286470, International TLB Research Institute, United States, 1996.
25. Marschner, P., Kandeler, E., Marschner, B., Structure and function of the soil microbial community in a longterm fertilizer experiment. *Soil Biol. Biochem.*, 35, 3, 453–461, 2003.
26. Samuel, A.L. and Ebenezer, A.O., Mineralization rates of soil forms of nitrogen, phosphorus, and potassium as affected by organomineral fertilizer in sandy loam. *Adv. Agric.*, 2014, 1–5, https://doi.org/10.1155/2014/149209, 2014.
27. Bouranis, D.L., Theodoropoulos, A.G., Drossopoulos, J.B., Designing synthetic polymers as soil conditioners. *Commun. Soil Sci. Plant Anal.*, 26, 9&10, 1455—1480, 1995.
28. Martin, W.P., Status report on soil conditioning chemicals. *Soil Soc. Am. J.*, 17, 1, 1–9, 1953.
29. Hickman, J.S. and Whitney, D.A., *Soil conditioners*, Nort Central Regional Extension Publication 295, Kansas State University, Manhattan, Kansas, 2006.
30. Sheridan, R.C., Chemical fertilizers in southern agriculture. *Agric. Hist.*, 53, 1, 308–318, 1979.
31. Guedes, A.C., Amaro, H.M., Pinto, İ.S., Malcata, F.X., *Algal spent biomass – A pool of applications*, pp. 397–433, Elsevier Inc., Amsterdam, Netherlands, 2019.
32. Carter, O.G., The effect of chemical fertilizers on seedling establishment. *Aust. J. Exp. Agric. Anim. Husb.*, 7, 174–180, 1966.
33. Gellings, C.W. and Parmenter, K.E., *Energy Efficiency in fertilizer production and use*, pp. 122–136, Eolss Publishers Co. Ltd., Oxford, UK, 2009.
34. Loneragan, J.F. and Snowball, K., Calcium requirements of plants. *Aust. J. Agric. Res.*, 20, 465–478, 1969.
35. Simon, E.W., The symptoms of calcium deficiency in plants. *New Phytol.*, 80, 1–15, 1978.
36. Guo, W., Nazim, H., Lianf, Z., Yang, D., Magnesium deficiency in plants: An urgent problem. *Crop J.*, 4, 83–91, 2016.
37. Lewandowska, M. and Sirko, A., Recent advances in understanding plant response to sıulfur-deficiency stress. *Acta Biochim. Pol.*, 55, 457–471, 2008.
38. White, J.G. and Zazoski, R.J., Mapping soil micronutrients. *Field Crops Res.*, 60, 11–26, 1999.
39. Shuman, L.M., Micronutrient Fertilizers. *J. Crop Prod.*, 1, 2, 165–195, 1998.
40. Millán, A.F.L., Grusak, M.A., Abadía, A., Abadía, J., Iron deficiency in plants: an insight from proteomic approaches. *Front. Plant. Sci.*, 4, Article 254, 1–7, 2013.
41. McHargue, J.S., The role of manganese in plants 1. *J. Am. Chem. Soc.*, 44, 7, 1592–1598, 1922.

42. Alloway, B.J., *Zinc in soils and crop nutrition*, pp. 1–135, International Fertilizer Industry Association, and Zinc International Zinc Association, Brussels, Belgium, 2008.
43. Pendias, A.K., *Trace elements in soils and plants*, CRC Press LLC, Florida, 2001.
44. Loomis, W.D. and Durst, R.W., Chemistry and biology of boron. *Biofactors*, 3, 4, 229–239, 1992.
45. Reed, H.S., A physiological study of boron deficiency in plants. *Hilgardia*, 17, 11, 377–411, 1947.

8

Heavy Metal Chemistry in Soils

Sapna Nehra[1], Rekha Sharma[2] and Dinesh Kumar[3]*[

[1]*Department of Chemistry, Dr. K. N. Modi University, Newai, Rajasthan, India*
[2]*Department of Chemistry, Banasthali Vidyapith, Banasthali, Rajasthan, India*
[3]*School of Chemical Sciences, Central University of Gujarat, Gandhinagar, India*

Abstract
Globally, soil contamination by heavy metals has become an environmental issue. Soil is a functional constituent of the biosphere which unprotected from the accumulation of heavy metals in higher concentrations like the arsenic, cadmium, lead, mercury, cobalt, copper, nickel, zinc, selenium, and many more. The accumulation of heavy metals in the soil can be because of natural and anthropogenic activities. The increasing concentration of the heavy metals shows the adverse effect on the living organism and the agriculture fields in the ecosystem. To remove the solemn ecological difficulties and danger to human health attributed by the heavy metals in the soil makes very important its removal by the employment of chemical methods. In the past few decades, cropland ecosystem pollution has been largely increased in various countries. So, various kinds of techniques were utilized for the treatment of heavy metals in the soil as extraction and immobilization. This chapter discusses the various used approaches for the mitigation of the heavy metals from the soil.

Keywords: Redundancy analysis, soil organic matter, hazard quotients, multivariate regression tree, contamination factor

8.1 Introduction

The rapid growth of the population and urbanization increased the amount of heavy metal in the water system. The generation of batteries, thin-film coating, release by-product, and automotive colors worsens the quality

**Corresponding author*: dinesh.kumar@cug.ac.in

Inamuddin, Mohd Imran Ahamed, Rajender Boddula and Tariq Altalhi (eds.) Applied Soil Chemistry, (149–164) © 2021 Scrivener Publishing LLC

of water. Fast industrialization because of the development of batteries, automotive stains, steel, and thin-film coating with inappropriate capacity and elimination of waste, including electrical and electronic equipment, produces the toxic wastewater [1, 2]. Heavy metals cause the number of diseases. Lead causes the major dysfunction in the liver, kidneys, and reproductive system. Cadmium causes destruction of the kidney, RBC, testicular tissue, and osteoporosis and enhances the blood pressure. Copper also causes diarrhea, nausea, stomach cramps, vomiting, and even death [3]. The assimilation of zinc in soil attributes the irregular production of wheat, yellow coloration in leaves, and development dwarf plants. A large amount of zinc degrades the quality of soil and dropping the quantity of the bacteria in the soil [4–7]. Nowadays, the purification of the soil has been done by the utilization of the mechanical and chemical methods like evaporation, dispersion, washing and burying [8]. Several newly chemical remediation methods take short time intervals and small distribution of the hazardous constituents, which increase the soil washing. Further, the methods include the different categories as electrochemical, electrodialysis, evaporation recovery, chemical precipitation, circulation-enhanced electro-kinetics, solvent extraction, reverse membrane osmosis, microbial electrochemical system, metal redox treatment, ultrafiltration, biological filtration and phytoremediation, ion exchange, and adsorption [9–11]. Low-cost chelating agents have extensively known and offer an appropriate effective method for the remediation of toxic heavy metals from the soil comprises the sewage sludge biochar, agricultural wastes, biomass, clays, and coconut shell [12, 13]. Although, the heavy metals exhibited the tendency to strongly bind with organic matter present in the soil. So, more research focuses on the utilization of the chemical agents to minimize the concentration of the heavy metals found in the soil. The uptake of heavy metal by the crops is also diminished because of lesser concentration in soil [14, 15]. Various reducing agents as remediating carries have been utilized like ascorbic acid, hydroxylamine hydrochloride, sodium disulfate, and sodium oxalate. The oxidants are persulfate, hydrogen peroxide, surfactants, acids, and bases that increase the elimination of heavy metals from the soil, which shows the adverse effect on microorganisms found in the soil and also reduces the fertility of the soil. Therefore, to remove these limitations, researchers focus on the adsorbents that have been derived from biomass and biosorption like biomaterials, a further sum of the plant matter, agricultural waste, and marine organisms [16, 17]. Mainly, the solidification, as well as the stabilization, is thought out to be the most effective method for remediation in the soil [18]. The effective solidification of heavy metals occurred over the adsorbent surface, the biodegradation of the soil, as well as the mobility of the metal ions, which generated secondary pollution [19]. The other major issue is the existence of the various metallic cations like sodium, potassium, and calcium, many more, which diminished the effectiveness of the adsorbent for the elimination of the hazardous heavy metal found in the soil. Meanwhile, cleaning soil is a very significant strategy to target the heavy metals by using the cleaning liquor, and the effect over the soil could be noticed. In previous studies, the cleaning of soil has been performed with the involvement of the surfactants, dynamic redox, alkali solutions, minerals, inorganic and organic acids, and chelating agents like EDTA, [S, S]-EDDS, phosphoric acid, and oxalic acid, which breaks down the remaining heavy metal in the soil [20]. The effect of ultrasound over the acidic medium containing soil is a very effective method for the remediation of the contaminated heavy metals found in the soil [21]. Mostly used methods seem to be not much as efficient for the soil because it shows very adverse impact on the fertile microbes present in the soil. But it has shown the cleaning of soil by the addition of the chelating agents the safe level [22]. The chemical, and the physical nature of the soil and their corresponding present microbes, has been shown the illustrative assessment of the removal process [23]. Other than these substrates, biochar also has been utilized to eliminate heavy metals from the soils [24–27]. It is a very cost-effective material for the mitigation of the heavy metals and controls the pollutants, which result in enhancing the soil quality in terms of nutrient recharge, water loading capacity, alkalinity, and better microbial as well as antibacterial property [28–33].

8.2 Outline of the Heavy Metal Chemistry in Soil

Zhao *et al.* prepared the decomposable LBPAA composite with the help of a chelating agent for the mitigation of the heavy metals, including the copper, zinc, cadmium, and lead found in the soil. By this biodegradable composite, the abstraction and ion transference of heavy metal ions enhanced the elimination rate of toxic soil at the time of water leaching. Afterward, five times consecutive washing the contaminated soil, the concentration of heavy metals that comprise lead, copper, zinc, and cadmium was diminished up to 17.95%, 22.57%, 52.60%, and 13.63%, correspondingly, and obtained 1.31-, 2.39-, 5.04-, and 5.04-fold more dominating result than cleaning with the deionized water. The composite remained the ability of the sequesters for the cadmium and the lead from the toxic soil and easily released into the eluent. Therefore, this present work provides a very effective insight related to lignin as a safe, eco-friendly, and sustainable mitigation method for the hazardous soil exhibited contaminate like heavy metals [34].

Xue *et al.* showed the effect of sewage water irrigation and the groundwater into the soil in the Northern China region and found the higher concentration of the heavy metals in the sewage irrigation regions. In the Northern China region, they have given sewage treatment to the soil to reduce the consumption of the groundwater. The observations were carried out in that farmland area, which situated at Fuhe River, where continuous grains of maize and wheat were treated with sewage irrigation. Further, the outcome results attributed to the higher accumulation of the cadmium, lead, and mercury over the topsoil compare to the groundwater irrigated water. They found 48% of cadmium concentration in the sewage water treated soil samples, which beyond the limit of 0.6 mg/kg, recommended by Chinese safety limitation. While a small amount of cadmium was stored, which affects the minimum to human health. Alternatively, the lead amount in the sewage-irrigated soil was found below the recommended limit, but around 24% concentration present in the wheat grain sample attributes exceeded the limit set by the Chinese national safety board. For long-duration exposure of crop grains as well as the soil does not increase the amount of other heavy metal like arsenic, chromium, and nickel. THQ was checked to determine the toxicity of the heavy metal-containing edible crop grains toward human health. Overall, they reported the THQ level of wheat grains samples over 1.0 regarding sewage irrigation as a good smelter affected area. Here, arsenic is most responsible for enhancing the THQ level. So, the assimilation of the cadmium, lead, mercury, and arsenic in the soils and grains that were grown in that area of swage irrigation should be assessed to confirm food safety and security [35].

Huang *et al.* summarized the heavy metal contamination in the china region. The accumulation of the eight kinds of heavy metals was analyzed in the reported study like cadmium, chromium, lead, mercury, arsenic, copper, nickel, and zinc. The quantity of cadmium and mercury was enhanced compared to previous concentration, and the rest of six heavy metals do not increase as much concentration. Smaller extent pollution was reported in the farmland area, which has very different from the other anthropogenic emitted pollutants. Although the production of cadmium and mercury continuously accumulated from the mining smelting and industrial areas. But, the growth, which gets slower since 2012, could be due to lesser utilization of the coals, agrochemicals, and nonferrous metals. The quantity of the heavy metals mostly found maximum in the southwest and south coastal areas, but minimum in the northwest regions.

It demonstrated higher content in the vegetable and rice arenas compare to the other used land. This report estimated the serious causing factors of soil pollution in the farmland regions at the national level. This assessment will support the controlling and management approaches of newly heavy metal in China [36].

Xiao *et al.* described how large the content of the heavy metals in the soil affects crop production and the quality of the soil. Here, the distribution of the heavy metals in the rice and vegetable arenas. It supplied at the nearby place of electroplating industry and wastewater for the irrigation is demonstrated. The remarkable accessibility was checked through the SPF of heavy metals, comprises the reducible, oxidizable fractions, and acid-soluble obtained from the BCR consecutive abstraction process. For the testing width of 0 to 20 cm, topsoil tasters were prepared. Overall, it also assessed the SPF value of the heavy metals and their corresponding physicochemical features of the soil were as well. During the analysis, multi statistical variables, including PCA and RDA, were utilized. The obtained results displayed the total, and SPF quantity of toxic metals in the soil tasters reduced up to $P < 0.05$ as long-distance soil from the industry determined the taster distance is the serious factor in evaluating the influence of heavy metals in soil. As per PCA, the collected soil samples were dispersed over the PCA line, which shows mankind's influences, evidenced by the alteration in the soil samples obtained from the irrigation with electroplating wastewater. Redundancy analysis as well as the stepwise regression specified that soil manganese oxide, amorphous iron oxides, pH, and sediment quantity might be predicting the alteration up to 68.8% and 43.5% in SPF concentration of rice and vegetable growing soil. The probable accessibility of heavy metal positively depends upon manganese oxide but negatively linked with iron oxide content present in the soil. It means that manganese oxide directly enhances the availability of heavy metals, but iron oxide diminished its accessibility [37].

Wang *et al.* investigated the impact of 5% wood char, rice straw, Chinese walnut shell biochar, bamboo biochar on growth of heavy metals gathering in the Moso bamboo, and soil physiochemical features and dissolution of heavy metals in soil. The obtained outcomes confirmed that the dry weight of Mosso bamboo has highly risen with the utilization of 5% wood biochar, rice biochar, and Chinese walnut shell biochar despite bamboo biochar. Among all this, 5% straw biochar is more significant to raise the plant biomass by enhancing the 157%, 113%, and 111% in leaves, roots, and stems of Moso bamboo. In terms of storage of heavy metals, the importance of 5% Chinese walnut shell, bamboo, straw biochar, and Chinese walnut diminished copper loading in roots by 15%, 35%, and 26%, correspondingly. Biochars played a vital role in decreasing the dissolution of heavy soil metals with more elimination of 3658.91 and 10.59 mg/kg of copper and lead [38].

Cheng *et al.* investigated the reflectance spectra to determine the amount of the cadmium, lead, arsenic, chromium, copper, and zinc in suburban soils and also equate the capabilities of diverse spectral pretreatments. With the help of visible and near-infrared reflectance spectra, the amount of metal is examined. To regulate the VNIR model, partial least-square regression was utilized. The outcomes displayed in this model attributed the accurate concentration of the chromium, arsenic, and cadmium with the 2.70, 1.81, and 1.63 of RPD value. Meanwhile, they found less accuracy for the estimation of lead, copper, and zinc concertation with 0.70 to 1.03 RPD values. The cadmium amount is particularly assessed and nearly laid to the SOM. Chromium and arsenic group (II) amount can be assessed by the VNIR model on behalf of iron, lead, copper, and zinc concentrations. This spectroscopic model associated with PLSR is a significant process that examined the concentration of some heavy metal pollution in suburban soils [39].

Mao *et al.* determined the transference and gathering ways of different heavy metals and its geochemical behavior of crop growing fields of rice near Yangtze River Delta. Initially, soil exhibited the large place of accumulation of heavy metal, and through this, it passes into the food chain. The experiment results show that the concentration of the heavy metal reduced by enhancing soil pH. But, rice shoots stored many heavy metals rapidly in the presence of less pH of soil. The HQ of heavy metals such as lead and arsenic shows the adverse effect to human health. Subsequently, the cancer was more susceptible when the concentration of the cadmium and arsenic was found approximately 76% and 15.7%. The reduced pH condition of the soil was produced a toxic and nontoxic effect on human health. With the help of exponential alteration of grain and soil metals, the value of HQ estimated the rice plants that used the regression curves. It is assumed that soil containing high amount of heavy metal in rice grown nontoxic and cancer risk could be assed in particular levels after harvesting like arsenic, lead, cadmium, copper < 20, < 100, $< 0.07–0.68$, $< 7.56–30.87$, and arsenic, cadmium, chromium $< 20 < 4 < 200$ mg/kg, respectively [40].

Jiang *et al.* described a heavy metal at e-waste sites in Nigeria, an obtained distinction in the distribution of the comparing surface area soil and subsoils. Dominantly, the Firmicutes, Acidobacteria Planctomycetes, and Proteobacteria found in the native soil microbial societies and seemed to very large discernment in bacterial taxonomic composition among the polluted and unpolluted areas. The richness of a large variety of bacterial taxa altered with heavy metal pollution to a certain level. The MRT analysis described the chief environment factor of soil like 31% of soil texture and 14% of organic carbon, which affects the bacterial composition diversity, and 32% of pH and 14% of soil texture were affected the microbial taxonomy. The present study outcomes showed that the texture of the soil is one of the keen factors to determine the composition of soil bacteria, and the diversity, compared to heavy metals at e-waste sites, which polluted [41].

Fei *et al.* performed an accurate rating of the latitudinal distribution of heavy metal contaminants. The utilization of supplementary soil features data, which can enhance the mapping, BME, and GWR model utilized to examine the Shanghai City soils (China). These both used models have been more efficient method than the previous traditional methods because of some advantages. It assimilates heavy metal quantities and supplementary data from the sound theoretical basis, and better perform both accuracy calculation and implementation flexibility and assimilation of multiple data sources. Based on BME-GWR, the quantity of heavy metal maps was developed and reported the more amount of heavy metal that includes arsenic, chromium, and lead in Shanghai, while a higher concentration of cadmium was reported in the northwestern area of Shanghai. They found the organic carbon and the pH with heavy metal in the city soil. Here, particularly the cadmium and chromium pollution are a consequence of the agricultural activities and the natural sources, although lead and arsenic have a different source of causing pollution [42].

In the work of Lv *et al.*, APCS/MLR and PMF used over cumulative data comprise 10 heavy metals in 300 surface soil models. Robust geostatistics was used to define and match the points that were obtained from the two receptor models. The APCS/MLR and PMF contributed similar three features equivalent with aids, while APCS/MLR exhibited undesirable an unrecognized feature. So, PMF showed the optimal non-negativity out comes for source distribution. Through the PMF and robust geostatistics, three foundations of 10 heavy metals in Guangrao were reported: arsenic, cobalt, chromium, copper, manganese, nickel, zinc, and in a small amount of mercury, lead, and cadmium created from the natural source. Through the gathering feature, heavy metal shows the uniform dispersion with the parent material map. Around 43.1% and 13.2% of mercury and lead were found in atmospheric deposition through human activities in the urban areas. Approximately, 29.6% of the cadmium comes in the atmosphere through the agricultural activities, and it generated the hotspots via the longitudinal dispersion of that soil, which produced the vegetables. Totally, usual sources, atmospheric accumulation by human activities, and agricultural activity contributed to the corresponding 10 heavy metal quantity as

81.1%, 7.3%, and 11.6%, respectively. Therefore, the receptor model united robust geostatistics could provide efficient data to assess the foundation and the distribution of heavy metals in soils [43].

Ayoubi *et al.* utilized the magnetic parameters and selected soil features to assess the number of heavy metals such as chromium, cobalt, iron, copper, nickel, manganese, and zinc in soils. These heavy metals were eliminated from the rocks present in Kurdistan province, western Iran. They collected 105 samples from the rocks and their surrounded field area. Magnetization was checked at two frequencies; it demonstrated the quantity of nominated metal and physiochemical features of the soil. From gabbro rocks, eliminated soil has exhibited the maximum amount of heavy metals like cobalt, nickel, iron, manganese, and copper. While porphyritic-granite and spilite basalt rocks produced soil that exhibited the zinc and chromium higher concentration. Magnetic susceptibility tested the change ability in the amount of heavy metals in soil, and their corresponding rocks was 78%, 74%, 77%, 72%, 75%, 68%, and 69% of the total of zinc, copper, nickel, iron, cobalt, and chromium quantity in the collected soils. Physicochemical features not much predict the quantity of heavy metals. So, it assumes that the magnetic susceptibility is the rapid, cost-effective, and non-destructive method, which describes the heavy metal concentration in the natural ecosystem [44].

Khalid *et al.* selected five different sites for the Faisalabad-Sargodha road (FSR), Pakistan, to assess the concentration of the heavy metals comprise lead, nickel, and zinc cadmium by using the Xanthium stramonium. They selected the leaf and shoot parts with soil samples to determine the storage tendency of these metals in it. For the estimation of the metabolism of the X. strumarium content of amino acid, photosynthetic pigments and the antioxidant property were checked. In the selected plant in their roots and shoot parts, the heavy metal concentration was assessed and compared with the control. Overall, the plant leave obtained the cadmium, lead, zinc, and nickel concentration from the soil and average concentration was 0.27 ± 0.01, 3.33 ± 0.16, 54.5 ± 1.02, and 5.85 ± 0.11 and 0.25 ± 0.24, 3.38 ± 0.29, 118.7 ± 1.04, and 4.89 ± 0.12 mg/kg, respectively. The smaller reduction was observed in the photosynthetic pigments in the plants, which grows in the high concentration heavy metal soil while maximum antioxidant activity. A remarkable point of heavy metals was noted, which shows the relationships with the laves and the soils. This study concluded that the X. strumarium acts as strong phytoremediator of metals in roadsides of the polluted area [45].

Kumar *et al.* reviewed the status of the presence of heavy metals in the soil of India from 1991 to 2018, which shows the zinc and lead presence above the recommended limit. The concentration limit was set for zinc and lead was 22.1, 13.1, 200, 80, and 300 μg/g, respectively, by Indian, Canada, Swedish, and Poland guidelines. The average value for the arsenic and copper in the roadside soils was very higher. In the Indian soils, the concentration was found to be for various heavy metals such as iron, manganese, zinc, copper, chromium, arsenic, nickel, lead, and copper as 23774.84, 872.54, 359.94, 183.67, 161.42, 148.70, 112.41, 61.87, 37.63, and 14.16 μg/g, correspondingly. To evaluate the various soils and displayed a diverse source of the metals, cluster and factor analysis was utilized. The CF, EF, and Cp confirmed the chief pollutants like cadmium and arsenic. The outcomes of the ecological risk presented the cadmium is one of the main pollutants found in Indian soil [46].

Vega *et al.* studied the mining effect on the environment of degradation. Mining formed a large extent of waste, which limits its physical, chemical, and biological limitations. The maximum concentration of the heavy metals was present in mining soils because it exhibited the metallic constituents. They collected and characterized the 25 soils from Touro and Meriama that is copper and lignite mine situated in Galicia (Spain). The DTPA approach was utilized to extract the dissolved amount of several metals like cadmium, chromium, copper, nickel, lead, and zinc. The dissolving amount of heavy metals and DTPA contents has very minimum in the soil, excluding the copper dispersible amount in the soil from copper mine spoils. The other metals such as chromium, nickel, and zinc eliminated from the parental substance of these soils. The quantity of the heavy metals in the Meirama mine soils, also gained from the fertilizer as well as the animal dung. Some other materials were also responsible for reducing the heavy metal accessibility, which comprises the manganese oxides, iron oxide, humified organic matter, and clay minerals as gibbsite, smectite, goethite chlorite [47].

Li *et al.* employed the chlorination and thermal strategy for the elimination of heavy metals from the polluted soils at the industrial scale. They collected five different heavy metal-containing soils and treated in the heated rotary kiln exhibited 17m length, 4m diameter, and rotating speed as 0.60 rpm. The given temperature was 950 °C and the added $CaCl_2$ for the chlorination method to the soil during this treatment. Furthermore, these treatments show the good potential to eliminate the lead and cadmium. Besides, these also removed the copper and zinc in a smaller extent while not much as efficient for the chromium and nickel. So, this treatment remarkably reduced the bio accessibility of heavy metals in soils, especially lead and cadmium. The consecutive abstraction exposed with thermal handling declined the carbonate, iron, manganese oxide, and organic matter. It also enhanced the residual amount of heavy metal, which revealed the chlorination and thermal treatment attributed to the immobilization of heavy metals. The experimental outcomes of this study show the potential applicability and feasibility [48].

Bashir *et al.* demonstrated the particular toxicity of the cadmium ion in the agricultural crop fields. It can reduce transfer of the cadmium in the polluted soil through the alteration of soil, which decreased the cadmium passing up to plants from the soil. This study used the sepiolite to check the dissolution and the gathering of the cadmium in the spinach tissue. The pH of the soil, cadmium content, and its storage adsorption mechanism also are investigated in spinach tissue. The addition of the sepiolite from 1 to 5% in the soil easily enhanced its pH from 0.3 to 1.0 units. Correspondingly, cadmium quantity in acid dissolution was reduced around 42.8% and rose the residual amount around 35.8% with a 5% rate. Besides, the substantial decrement in the cadmium loading through the spinach shoots was 26.2%, and the root part was 30.6% at a 5% rate. The higher adsorption tendency was 37.35 mg/g was obtained with a 5% rate. Several analytic tools were utilized as XRD, FTIR, and SEM, to check the potential of the sepiolite for the cadmium-contaminated soil renewal and thus declined its phytoavailability in contaminated soil to improve food security tasks [49].

Wan *et al.* performed a field experiment to determine the use of chicken or swine dung on the lead, cadmium, chromium, and arsenic accessibility, content amount, and its storage in the soil and loading through paddy plant. The experimental outcomes displayed the chicken and swine dung remarkably decrease the cadmium and lead traces in the paddy grains by 7.8 to 79.3% and 7.2 to 59.4%, correspondingly. The utilization of the dung considerably enhanced the DTPA-extractable arsenic and transferrable arsenic contents in the soil. Alternatively, it shows an inadequate impact over the arsenic storage in the rice crop. There is no big discrimination found in the accessibility and storage in the rice grain for the chromium. So, cattle manure can be employed as soil alterations to reduces the content of cadmium and lead in the paddy crop grain [50].

8.3 Conclusions

The present chapter comprises a number of approaches that have been performed to control the overabundance of heavy metals present in the soils, which further used for the agriculture purpose and largely affects the food chain. At the same time, some of the benefits and disadvantages found in the methods used for analysis and assessment of heavy metals in soil samples are discussed. Based on its results confirmed, the main heavy metal higher concentration was investigated in all reports rice crop field soil. The plethora of heavy metals is everywhere, and there is not the same concentration everywhere it changes the site to the site. This chapter will help readers to understand the used techniques in the past few years to estimate the heavy metal content and their mitigation process.

8.4 Abbreviations

EDDS = [S,S]-ethylene-diamine-disuccinic- acid
EDTA = Ethylenediaminetetraacetic acid
LBPAA = Lignin based poly(acrylic acid)
THQ = Target hazard quotient
SPF = Secondary phase fraction
PCA = Principal component analysis
RDA = Redundancy analysis
RPD = Ratio of the performance to deviation
SOM = Soil organic matter
HQ = Hazard quotients
TF = Transfer factor
e-waste = Electronic waste
MRT = Multivariate regression tree
BME = Bayesian maximum entropy
GWR = Geographically weighted regression
APCS/MLR = Absolute principal component score/multiple linear regression
PMF = Positive matrix factorization
PGPR = Plant growth promoting rhizobacteria
CF = Contamination factor
EF = Enrichment factor
Cp = Potential contamination index
DTPA = Diethylene triamine pentaacetate
XRD = X-ray powder diffraction
FTIR = Fourier-transform infrared
SEM = Scanning electron microscope

Acknowledgment

Dinesh Kumar is thankful to DST, New Delhi, for financial support to this work (sanctioned vide project Sanction Order F. No. DST/TM/WTI/WIC/2K17/124(C).

References

1. Sharma, B., Singh, S., Siddiqi, N.J., Biomedical implications of heavy metals induced imbalances in redox systems. *Biomed. Res. Int.*, 2014, 640754, 2014.
2. Ghnaya, T., Mnassri, M., Ghabriche, R., Wali, M., Poschenrieder, C., Lutts, S., Abdelly, C., Nodulation by Sinorhizobium meliloti originated from a mining soil alleviates Cd toxicity and increases Cd-phytoextraction in Medicago sativa L. *Front. Plant Sci.*, 6, 863, 2015.
3. Zare, E.N., Motahari, A., Sillanpää, M., Nanoadsorbents based on conducting polymer nanocomposites with main focus on polyaniline and its derivatives for removal of heavy metal ions/dyes: a review. *Environ. Res.*, 162, 173, 2018.
4. Lu, K., Yang, X., Gielen, G., Bolan, N., Ok, Y.S., Niazi, N.K., Xu, S., Yuan, G., Chen, X., Zhang, X., Liu, D., Song, Z., Liu, X., Wang, H., Effect of bamboo and rice straw biochars on the mobility and redistribution of heavy metals (Cd, Cu, Pb and Zn) in contaminated soil. *J. Environ. Manage.*, 186, 285, 2017.
5. Liang, J., Yang, Z., Tang, L., Zeng, G., Yu, M., Li, X., Wu, H., Qian, Y., Li, X., Luo, Y., Changes in heavy metal mobility and availability from contaminated wetland soil remediated with combined biochar-compost. *Chemosphere*, 181, 281, 2017.
6. Zhou, R., Liu, X., Luo, L., Zhou, Y., Wei, J., Chen, A., Tang, L., Wu, H., Deng, Y., Zhang, F., Wang, Y., Remediation of Cu, Pb, Zn and Cd-contaminated agricultural soil using a combined red mud and compost amendment. *Int. Biodeterior. Biodegrad.*, 118, 73, 2017.
7. Feng, C., Zhang, S., Li, L., Wang, G., Xu, X., Li, T., Zhong, Q., Feasibility of four wastes to remove heavy metals from contaminated soils. *J. Environ. Manage.*, 212, 258, 2018.
8. Das, N. and Chandran, P., Microbial degradation of petroleum hydrocarbon contaminants: an overview. *Biotechnol. Res. Int.*, 2011, 941810, 2011.
9. Zhang, Y., Zhao, J., Jiang, Z., Shan, D., Lu, Y., Biosorption of Fe(II) and Mn(II) ions from aqueous solution by rice husk ash. *Biomed. Res. Int.*, 2014, 973095, 2014.
10. Chang, J.H., Wang, Y.L., Shen, S.Y., A specific configuration of circulation-enhanced electrokinetics (CEEK) to remediate real-site Cd and Pb contaminated soils. *J. Hazard. Mater.*, 359, 408, 2018.

11. Wei, H., Huang, M., Quan, G., Zhang, J., Liu, Z., Ma, R., Turn bane into a boon: application of invasive plant species to remedy soil cadmium contamination. *Chemosphere*, 210, 1013, 2018.
12. Gonzaga, M.I.S.C., Mackowiak., A., Quintão de Almeida., A., Wisniewski., D., Figueiredo de Souza., I., da Silva Lima., A., Nascimento de Jesus, Assessing biochar applications and repeated Brassica juncea L. production cycles to remediate Cu contaminated soil. *Chemosphere*, 201, 278, 2018.
13. Lakouraj, M.M., Mojerlou, F., Zare, E.N., Nanogel and superparamagnetic nanocomposite based on sodium alginate for sorption of heavy metal ions. *Carbohydr. Polym.*, 106, 34, 2014.
14. Guo, X., Zhao, G., Zhang, G., He, Q., Wei, Z., Zheng, W., Qian, T., Wu, Q., Effect of mixed chelators of EDTA, GLDA, and citric acid on bioavailability of residual heavy metals in soils and soil properties. *Chemosphere*, 209, 776, 2018.
15. Rehman, M.Z.U., Rizwan, M., Hussain, A., Saqib, M., Ali, S., Sohail, M.I., Shafiq, M., Hafeez, F., Alleviation of cadmium (Cd) toxicity and minimizing its uptake in wheat (Triticum aestivum) by using organic carbon sources in Cd-spiked soil. *Environ. Pollut.*, 241, 557, 2018.
16. Rehman, M.Z.U., Rizwan, M., Ali, S., Ok, Y.S., Ishaque, W., Saifullah, Nawaz, M.F., Akmal, F., Waqar, M., Remediation of heavy metal contaminated soils by using Solanum nigrum: a review. *Ecotoxicol. Environ. Saf.*, 143, 236, 2017.
17. Mahfoudhi, N. and Boufi, S., Nanocellulose as a novel nanostructured adsorbent for environmental remediation: a review. *Cellulose*, 24, 1171, 2017.
18. Taubert, M., Stöckel, S., Geesink, P., Girnus, S., Jehmlich, N., von Bergen, M., Rösch, P., Popp, J., Küsel, K., Tracking active groundwater microbes with D_2O labeling to understand their ecosystem function. *Environ. Microbiol.*, 20, 369, 2018.
19. Ma, Y., Sun, Y., Fu, Y., Fang, G., Yan, X., Guo, Z., Swelling behaviors of porous lignin-based poly (acrylic acid). *Chemosphere*, 163, 610, 2016.
20. Beiyuan, J., Tsang, D.C.W., Valix, M., Zhang, W., Yang, X., Ok, Y.S., Li, X.D., Selective dissolution followed by EDDS washing of an e-waste contaminated soil: extraction efficiency, fate of residual metals, and impact on soil environment. *Chemosphere*, 166, 489, 2017.
21. Park, B. and Son, Y., Ultrasonic and mechanical soil washing processes for the removal of heavy metals from soils. *Ultrason. Sonochem.*, 35, 640, 2017.
22. Beiyuan, J., Awad, Y.M., Beckers, F., Tsang, D.C.W., Ok, Y.S., Rinklebe, J., Mobility and phytoavailability of As and Pb in a contaminated soil using pine sawdust biochar under systematic change of redox conditions. *Chemosphere*, 178, 110, 2017.
23. Im, J., Yang, K., Jho, E.H., Nam, K., Effect of different soil washing solutions on bioavailability of residual arsenic in soils and soil properties. *Chemosphere*, 138, 253, 2015.
24. Lee, A.L.Z., Voo, Z.X., Chin, W., Ono, R.J., Yang, C., Gao, S., Hedrick, J.L., Yang, Y.Y., Injectable coacervate hydrogel for delivery of anticancer

drug-loaded nanoparticles *in vivo*. *ACS Appl. Mater. Interfaces*, 10, 13274, 2018.
25. Zare, E.N., Lakouraj, M.M., Kasirian, N., Development of effective nano-biosorbent based on poly m-phenylenediamine grafted dextrin for removal of Pb (II) and methylene blue from water. *Carbohydr. Polym.*, 201, 539, 2018.
26. Ma, Y., Lv, L., Guo, Y., Fu, Y., Shao, Q., Wu, T., Guo, S., Sun, K., Guo, X., Wujcik, E.K., Guo, Z., Porous lignin based poly (acrylic acid)/organo-montmorillonite nanocomposites: swelling behaviors and rapid removal of Pb (II) ions. *Polymer*, 128, 12, 2017.
27. Hughes, D.L., Afsar, A., Laventine, D.M., Shaw, E.J., Harwood, L.M., Hodson, M.E., Metal removal from soil leachates using DTPA-functionalised maghemite nanoparticles, a potential soil washing technology. *Chemosphere*, 209, 480, 2018.
28. Xue, W., Huang, D., Zeng, G., Wan, J., Cheng, M., Zhang, C., Hu, C., Li, J., Performance and toxicity assessment of nanoscale zero valent iron particles in the remediation of contaminated soil: a review. *Chemosphere*, 210, 1145, 2018.
29. Cao, M., Tu, S., Xiong, S., Zhou, H., Chen, J., Lu, X., EDDS enhanced PCB degradation and heavy metals stabilization in co-contaminated soils by ZVI under aerobic condition. *J. Hazard. Mater.*, 358, 265, 2018.
30. Lu, H.P., Li., Z.A., Gascó, G., Méndez, A., Shen, Y., Paz-Ferreiro, J., Use of magnetic biochars for the immobilization of heavy metals in a multi-contaminated soil. *Sci. Total Environ.*, 622–623, 892, 2018.
31. Tica, D., Udovic, M., Lestan, D., Immobilization of potentially toxic metals using different soil amendments. *Chemosphere*, 85, 577, 2011.
32. Ma, Y., Li, X., Mao, H., Wang, B., Wang, P., Remediation of hydrocarbon–heavy metal co-contaminated soil by electrokinetics combined with biostimulation. *Chem. Eng. J.*, 353, 410, 2018.
33. Chen, H.M., Zheng, C.R., Tu., C., Shen., Z.G., Chemical methods and phytoremediation of soil contaminated with heavy metals. *Chemosphere*, 41, 229, 2000.
34. Zhao, T., Zhang, K., Chen, J., Shi, X., Li, X., Ma, Y., Fang, G., Xu, S., Changes in heavy metal mobility and availability in contaminated wet-land soil remediated using lignin-based poly (acrylic acid). *J. Hazard. Mater.*, 368, 459, 2019.
35. Xue, P., Zhao, Q., Sun, H., Geng, L., Yang, Z., Liu, W., Characteristics of heavy metals in soils and grains of wheat and maize from farmland irrigated with sewage. *Environ. Sci. Pollut. Res.*, 26, 5554, 2019.
36. Huang, Y., Wang, L., Wang, W., Li, T., He, Z., Yang, X., Current status of agricultural soil pollution by heavy metals in China: A meta-analysis. *Sci. Total Environ.*, 651, 3034, 2019.
37. Xiao, L., Guan, D., Chen, Y., Dai, J., Ding, W., Peart, M.R., Zhang, C., Distribution and availability of heavy metals in soils near electroplating factories. *Environ. Sci. Pollut. Res.*, 26, 22596, 2019.

38. Wang, Y., Zhong, B., Shafi, M., Ma, J., Guo, J., Wu, J., Ye, Z., Liu, D., Jin, H., Effects of biochar on growth, and heavy metals accumulation of moso bamboo (Phyllostachy pubescens), soil physical properties, and heavy metals solubility in soil. *Chemosphere*, 219, 510, 2019.
39. Cheng, H., Shen, R., Chen, Y., Wan, Q., Shi, T., Wang, J., Wan, Y., Hong, Y., Li, X., Estimating heavy metal concentrations in suburban soils with reflectance spectroscopy. *Geoderma*, 336, 59, 2019.
40. Mao, C., Song, Y., Chen, L., Ji, J., Li, J., Yuan, X., Yang, Z., Ayoko, G.A., Frost, R.L., Theiss, F., Human health risks of heavy metals in paddy rice based on transfer characteristics of heavy metals from soil to rice. *Catena*, 175, 339, 2019.
41. Jiang, B., Adebayo, A., Jia, J., Xing, Y., Deng, S., Guo, L., Liang, Y., Zhang, D., Impacts of heavy metals and soil properties at a Nigerian e-waste site on soil microbial community. *J. Hazard. Mater.*, 362, 187, 201.
42. Fei, X., Christakos, G., Xiao, R., Ren, Z., Liu, Y., Lv, X., Improved heavy metal mapping and pollution source apportionment in Shanghai City soils using auxiliary information. *Sci. Total Environ.*, 661, 168, 2019.
43. Lv, J., Multivariate receptor models and robust geostatistics to estimate source apportionment of heavy metals in soils. *Environ. Pollut.*, 244, 72, 2019.
44. Ayoubi, S. and Karami, M., Pedotransfer functions for predicting heavy metals in natural soils using magnetic measures and soil properties. *J. Geochem. Explor.*, 197, 212, 2019.
45. Khalid, N., Noman, A., Aqeel, M., Masood, A., Tufail, A., Phytoremediation potential of Xanthium strumarium for heavy metals contaminated soils at roadsides. *Int. J. Environ. Sci. Technol.*, 16, 2091, 2019.
46. Kumar, V., Sharma, A., Kaur, P., Sidhu, G.P.S., Bali, A.S., Bhardwaj, R., Thukral, A.K., Cerda, A., Pollution assessment of heavy metals in soils of India and ecological risk assessment: A state-of-the-art. *Chemosphere*, 216, 449, 2019.
47. Vega, F.A., Covelo, E.F., Andrade, M.L., Marcet, P., Relationships between heavy metals content and soil properties in minesoils. *Anal. Chim. Acta*, 524, 141, 2004.
48. Li, J., Hashimoto, Y., Riya, S., Terada, A., Hou, H., Shibagaki, Y., Hosomi, M., Removal and immobilization of heavy metals in contaminated soils by chlorination and thermal treatment on an industrial-scale. *Chem. Eng. J.*, 359, 385, 2019.
49. Bashir, S., Ali, U., Shaaban, M., Gulshan, A.B., Iqbal, J., Khan, S., Husain, A., Ahmed, N., Mehmood, S., Kamran, M., Hu, H., Role of sepiolite for cadmium (Cd) polluted soil restoration and spinach growth in wastewater irrigated agricultural soil. *J. Environ. Manage.*, 258, 110020, 2020.
50. Wan, Y., Huang, Q., Wang, Q., Yu, Y., Su, D., Qiao, Y., Li, H., Accumulation and bioavailability of heavy metals in an acid soil and their uptake by paddy rice under continuous application of chicken and swine manure. *J. Hazard. Mater.*, 384, 121293, 2020.

9

Modeling of Pollutant Mobility in Soil

Jülide Hızal

Yalova University, Engineering Faculty, Chemical Engineering Department, Yalova-Turkey

Abstract

In this study, the importance and benefits of modeling studies about retaining and mobilization of pollutants throughout soil and soil components were explained and summarized by considering the studies performed in recent years. The surface complexation model, which mostly defines the interaction between heavy metals and active sites of soil, was evaluated as regards of used PC programs such as FITEQL, MINEQL, and MINTEQ. When inspecting heavy metal sorption on solid phases such as clays, clays mineral, and soils, the electrical double layer between aqueous metal ion and solid/soil surface has to be marked. Constant Capacitance Model (CCM), Diffuse Layer Model (DLM), Stern Model, Triple Layer Model (TLM), and Charge Distribution Multi-Site Complexation (CD-MUSIC) have been used for this purpose. The required parameters of each model are input while running PC programs. The programs determine the chemical equilibrium constants and concentrations of active sites from experimental data. Thus, it enables to calculate predicted values, which allows forecasting the behavior of pollutant in environment.

Keywords: Modeling, clay, soil, FITEQL, heavy metal

9.1 Introduction

Soil occurs as a result of weathering of rocks and accumulation of mineral matter which has transported by wind or stream. The weathering is defined as physical, chemical, or biological transformation of rocks by size reduction, degradation, transformation, and dissolution. As a result of

Email: hizalyucesoy@gmail.com

Inamuddin, Mohd Imran Ahamed, Rajender Boddula and Tariq Altalhi (eds.) Applied Soil Chemistry, (165–182) © 2021 Scrivener Publishing LLC

chemical weathering, the soil involves many constituents such as metalloids, alkaline chlorides, sulfides, oxides and hydroxides, silicates, sulfates, chromates, borates, carbonates, and organic compounds. The existence of organic matter in soil sourced from dead or alive organism as plants and/or animals [1]. The dead organisms undergo humification process and, consequently, converts to humic substances. The humic substances are primary consituents of natural organic matter (NOM) in environment. Especially, humic and fulvic acids, which are subdivisions of humic substance, may have carboxylic, phenolic, aliphatic, and aromatic groups depending the conditions of formation. Thus, humic and fulvic acids determine the chemical physicochemical properties of soils, such as aggregate stability, sorption/desorption of organic compounds, mobilization, and bioavailability of heavy metal cations, via these functional groups [2]. The primary metal oxides of soil are silica (SiO_2), alumina (Al_2O_3), and iron(III) oxide (Fe_2O_3). These metal oxides are involved in clays and clay minerals such as kaolinite, montmorillonite, zeolite, etc. Complex structures, formed between humic substances and clay minerals, significantly affect physical, chemical, and biological properties of soil and sediments [3]. The humic substance and clay minerals as being primary contents of soil allow retention of persistent pollutants.

The fate of toxic metals in environmental systems such as soil, sediment, surface, and groundwater is of great interest to scientists due to their advers effects on living organism. It is possible to forecast their mobilization by simulating and modeling the experimental results of heavy metal retention. Modeling studies are important as regards of forecasting environmental phenomena such as mobilization and/or accumulation of either organic or heavy metal pollutants, and interaction of pollutants throughout complex system combined with soil, surface water, and ground water. Additionally, modeling studies are useful for better understanding the whole reactions occured in an environmental system and also enable to optimize the system.

In this study, surface complexation modeling, which enables to determine the fate of organic or heavy metal pollutants in soil, was discussed by considering surface acidity, double layer models, the reactions between soil and pollutants, and in respect the used modeling programs.

9.2 Modeling of Heavy Metal Sorption onto Soil or Soil Components

The modeling studies mostly need the usage of personal computer (PC) programs. At the begining of modeling studies, general information about which kind of interactions would occur in a system, what the acidity constant of surface is, which double layer model would use to define the interface between soil and aqueous solution in which the pollution is, and the certain and/or proximate concentrations of primary components of system must be input to program. There are many studies about adsorption modeling using PC programs [4–8]. These programs enable modeling of metal and ligand adsorption on all kind of adsorbents and modeling of surface acidity of these natural and synthetic adsorbents. Basic reactions explaining surface complexation are associated with heavy metal, ligand, proton, and charge of surface sites mostly having hydroxyl groups. The modeling studies basicly differentiate each other by the usage of different models defining electrical double layer at liquid-solid interface and which use different electrostatic correction factors in application of equations [9]. According to Gu *et al.*, the FITEQL program allows application of four types of electrical double layer models, which are Constant Capacitance Model (CCM), Diffuse Layer Model (DLM), Triple Layer Model (TLM), in surface complexation modeling [10]. If the data were obtained at different ionic strenghts (in that case, it means inspection of the existence of outer-sphere complexation), the TLM is generally used for optimization. On the other hand, CCM is especially used in the optimization of the data obtained the experiments performed at constant ionic strengths which points out the formation of inner-sphere complexes, and it assumes that all binding to surface are inner-sphere complexes and that there is a correlation between the surface potential and the surface charge density [10]. The CCM and DLM involves formation of inner-sphere complexes, while the TLM involves formation of outer-sphere complexes in addition to inner-sphere complexes [7].

The primary retention mechanism of heavy metals through soil environment is sorption. Heavy metal and ligand sorptions onto soil may occur as a result of various interactions such as adsorption via surface complexation, ion exchange, electrostatic interaction, and hydrophobic interaction. There are many studies dealing with this subject [11–14]. Many PC programs as ProtoFit, MINTEQ, MINEQL, and FITEQL have been used for modeling sorption onto hydrated oxides. The widely used modeling PC program is FITEQL which was firstly created and used by John Westall in 1982. FITEQL program solves complex equilibrium systems by using non-linear least squares optimization and chemical equilibrium model, which is generally surface complexation model. At the end of many iterations, it minimizes the difference between experimental and predicted data. The impeccability of convergence is understood with the WSOS/DF parameter. The WSOS/DF is the weighted sum of square residual divided by the degree of freedom. Thus, the program determines the chemical equilibrium constants and concentrations of active sites from experimental data. FITEQL program also allows activity coefficient correction using the Davies equation.

For determining the formation constants of surface complexes and the concentration of surface hydroxyl groups for Cd(II) uptake on a silanol surface (SOH) of a clay, at first, the surface deprotonation should be considered as below:

$$\equiv SOH_2^+ \leftrightarrows \equiv SOH + H^+ \qquad \text{Log } K_{a1}$$

$$\equiv SOH \leftrightarrows \equiv SO^- + H^+ \qquad \text{Log } K_{a2}$$

The second step is the determination of dissolved cadmium species in aqueous solution such as Cd^{2+}, $CdOH^+$, and Cd_2OH^{3+}. $CdOH^+$, and Cd_2OH^{3+} are hydrolization product of Cd(II) ion [9]:

$$Cd^{2+} + H_2O \leftrightarrows CdOH^+ + H^+ \qquad \text{LogK}_1 = -9.97$$

$$2Cd^{2+} + H_2O \leftrightarrows Cd_2OH^{3+} + H^+ \qquad \text{LogK}_2 = -8.40$$

Silanol surface of a clay has two different active sites: one of them is edge site (SOH) and the other is permanent negatively charged surface (X_2^{2-}). The proposed reactions for Cd(II) sorption on silanol surfaces is written as

$$\equiv SOH + Cd^{2+} \leftrightarrows \equiv SOCd^+ + H^+ \qquad \text{Log } K_1$$

$$X_2^{2-} + Cd^{2+} \leftrightarrows CdX_2 \qquad \text{Log } K_2$$

The next step is to determine the electrical double layer model. DLM surface complexation model can be chosen for this sample due to needing minimum number of adjustable parameters. These input parameters were introduced in Table 9.1.

The inputting of all reactions into FITEQL "Screen II: Species" is introduced in Table 9.2. The input experimental data are pH values and corresponding adsorbed amount of Cd(II) ion obtained pH dependent experiments. When the program runs, as a result of particular number of iterations, which was previously input to the program, formation constants of $SOCd^+$ and CdX_2 species are calculated. The predicted adsorbed amount of Cd(II) at any pH can be yielded by using these calculated Log K values

Table 9.1 FITEQL input values for the electrical double layer model (first screen in FITEQL).

EDL Model	DLM
Specific surface area (m^2/g)	27.9
Concentration of suspended solid (g/L)	5
Concentration of z:z electrolyte	0.1 M
Charge of z:z electrolyte	1
Site densities (mol/L)	8.0×10^{-5}

Table 9.2 A typical screen in FITEQL (II. Species) for hydrated silica - Cd(II) system.

	Log K	PSI	Cd(II)	SOH	(X2)2−	Cdads.	H+
Cd(II)	0	0	1	0	0	0	0
CdOH+	−9.97	0	1	0	0	0	−1
(Cd2OH)3+	−8.40	0	2	0	0	0	−1
(SOH2)+	3.6*	1	0	1	0	0	1
SOH	0	0	0	1	0	0	0
SO−	−8.3*	−1	0	1	0	0	−1
SOCd+	2.8**	1	1	1	0	1	−1
(X2)2−	0	−2	0	0	1	0	0
CdX2	6**	0	1	0	1	1	0
H+	0	0	0	0	0	0	1
OH−	−13.78	0	0	0	0	0	−1

*: [15]; **: estimated value.

in another run. Thus, the conformity between experimental and predicted data can be seen on a graph or it can be calculated by any statistical method.

Hizal and Apak modeled the Cu^{2+} and Pb^{2+} adsorption onto clays by using single- and double-site binding models of the Langmuir approach, and "two-site, two pKa" model by using FITEQL3.2 PC program. In that study, two different surface sites were defined as two edge sites which are pH-dependent sites: silanol ($\equiv S_1OH$) and alumina ($\equiv S_1OH$). While establishing the model, these were assumed that the metal ions bind to surface as in their unhydrolyzed form, that they interact with the active surface sites via monodentate surface complexation, and that the DLM is chosen for defining double layer between surface and aqueous solution due to needing the minimum number of adjustable parameters. It was reported that TLM was a complex model which needs many parameters such as capacitance of outer-sphere and binding constants belonging adsorption of background ions onto surface [10]. All reactions occured in the system were explained by formation of binary surface-metal and surface-ligand, ternary surface-metal-ligand complexes in the absence and presence of humic acid, respectively. The model was run considering complex species derived from surface and metal ions and their formation constants [16]. The used experimental data in FITEQL modeling is the measured adsorbed amount of adsorbate as a function of pH. As a result, the program calculates the formation constants of surface complexes and the molar concentrations of active sites.

In same study, double-site modified Langmuir model was used in order to further model except FITEQL. It is assumed that the maximum adsorption capacity (τ) is equal to sum of the predicted maximum adsortion capacities ($\tau_{max,1}$ and $\tau_{max,2}$) on both silanol and alumina sites. Equation 9.1 shows the non-linear equation for double-site modified Langmuir model. The known parameters in this equation are C_H (calculated from pH of sytem), C_F (which is equilibrium concentration of adsorbate), and τ (which is calculated from the application of experimental data to single-site modified Langmuir model).

$$[M_{ads.}] = \tau = \frac{\tau_{max,1}}{K_{1,1}^{-1}(C_H/C_F)+1} + \frac{\tau_{max,2}}{K_{2,1}^{-1}(C_H/C_F)+1} \quad (9.1)$$

By assigning these predicted values in Equation 9.1, the non-linear equation systems are solved; thus, the corresponding equilibrium constants of the surface-metal complexes are found for each group of surface sites.

As a result of both modeling studies, it was declared by the authors that the easier Cu(II) locating sites (most likely silanol sites) were occupied first. Thus, the Cu(II)-surface complex formed with high binding constant. Unlike from Cu(II), Pb(II) bound alumina sites having stronger basic character. On the other hand, the heavy metal adsorption in the presence of humic acid was explained according to surface complexation model. The authors described the surface complexation considering different functional groups of humic acid having different acidity constant. Providing HL_1 and HL_3 symbolized the sites having carboxyl and phenol functional groups of humic acid, respectively, it was said that species having close pK_a such as S_1OH and HL_1 or S_2OH and HL_3 preferred to interact with each other. The FITEQL modeling with theoretical assumptions described well whole heavy metal and humic acid retention in all cases. In further studies on modeling of adsorption, Hizal and Apak investigated the Cd(II) adsorption, and the effect of humic acid existence on Cd(II) adsorption. Hereby, they modeled metal cation adsorption based upon binary and ternary complex formation between surface, metal and ligand, and electrostatic interaction between Cd(II) ions and permanently negative-charged sites of clays [9]. In that study, the third surface site was added to model as being permanent negatively charged site on tetrahedral silanol layer (NaX). It was assumed that unhydrolyzed heavy metal ions form monodentate inner-sphere binary surface complexes with pH-dependent sites in the absence of humic acid. In the presence of humic acid, the adsorptions of sole humic acid, unhydrolized Cd(II), and Cd(II)-humate complex took place simultaneously. According to authors, the observed Cd(II) adsorption is described by the fact that the formation of binary ($\equiv S_2OCd$) and ternary ($\equiv S_1OCdL_1$ and $\equiv S_2OCdL_3$) surface complexes together with CdX_2 formation depending the pH of solution. L_1 and L_2 show dissimilar functional groups of humic acid which become active at different pHs. As a result of application of triple-site approach combined with FITEQL, the experimental data and the theoretical data calculated from the model perfectly conformed each other [9]. Similarly, Schaller *et al.* used two-site diffuse-layer surface complexation models (DLMs) for Cd(II) adsorption onto kaolinite and quartz by assuming monodentate metal-surface complexation. The ionic strength dependent adsorption on kaolinite was explained by interactions of Cd(II) with both variable charge sites (edge sites) and permanent negatively charged sites [17].

Unlike from the adsorption on kaolinite, Cd(II) adsorption on quartz was defined by one-site DLM occuring on edge sites. While modeling, various species of Cd(II) such as hydroxyl and carbonate were taken into account as well as the Cd(II) precipitation-dissolution equilibria. The stability constants of each complexation reaction were used as input values. As a result, it was declared that the Cd(II) adsorption onto quartz had small dependency to ionic strength, while the adsorption onto kaolinite was strongly ionic strength dependent due to having permanent negatively charged surface. On the other hand, the DLM fitted well the adsorption on kaolinite but it hardly conformed the adsorption on quartz [17].

The uranyl ion adsorption was similarly explained by the formation of inner-sphere surface complexation with the kaolinite edge hydroxyl sites and outer-sphere complexation with the permanent charge sites, and using single-site modified Langmuir model [18]. According to authors, unhydrolyzed UO_2^{2+} ions form monodentate surface complex with alumina sites, and electrostatically interact with NaX permanent negatively charged sites especially at low pHs. It was also expressed that the SCM-DLM provided extrapolation of radionuclide retention at geochemical conditions and better understanding of mobilization of radionuclides in environmental systems. Thus, it could be easier to forecast the mobilization of U(VI) from nuclear waste into soil layers [18]. Using similar approach, competitive adsorptions of Cu(II), Cd(II), and Pb(II) onto kaolinite type clays and montmorillonite were explained by using surface complexation model comprising inner-sphere complexation with silica and alumina sites, and outer-sphere complexation with permanent negatively charged surface [19, 20].

The heavy metal adsorption onto edge sites, in which aluminol groups are involved, of oxide and clay minerals occurs as a result of inner-sphere surface complexation by monodentate binding, while the metal ion interact with the permanent negatively charged sites on basal surfaces which results in inner-sphere surface complexation [10]. Vithanage *et al.* defined the Sb adsorption on natural red earth by surface complexation model by optimizing with FITEQL. The diffuse double layer model was used. The dominant species of Sb(III) and Sb(V) were determined as $Sb(OH)_3$ and $Sb(OH)_6^-$, respectively. Consequently, the Sb(III) adsorption was explained by the authors by inner-sphere surface complexation, while the Sb(V) adsorption occurred as a result of outer-sphere surface complex. Simulating the results to real system, the authors interpreted that releasing of Sb(V) into environment at higher pHs than 4 would increase [21].

The Cr(VI) adsorptions onto illite, kaolinite, birnessite, and ferrihydrite were investigated by using quantitative thermodynamic-based mechanistic surface complexation models (QTMSCM). The experiments were carried out at different pH, sorbate concentrations, ionic strength, and suspension concentration, and the surface complexation was proved by spectroscopic methods such as Fourier transform infrared spectroscopy (FTIR), X-ray photoelectron spectroscopy (XPS), and X-ray absorption spectroscopy (XAS). The dissolution of the solid phases was taken into account in surface complexation modeling. Combined SCM and spectroscopic approach well described the Cr(VI) retention in soil. On the other hand, it was reported that the QTMSCM had greater prediction for real soil systems. According to Stern Model, $H^+$and OH^- ions bind to surface via inner-sphere complexation, and the other ions bind via outer-sphere complexation. The titration data were evaluated by the aid of ProtoFit and FITEQL4.0 program, and the calculated values such as surface acidity constants, counter-ion binding constants, and the total number of surface sites were used as input values in modeling of surface complexation in FITEQL 4.0 running. At last, stability constants of each assumed reaction were used in Visual MINTEQ which is thermodynamic database program; thus, final DLM and/or TLM was achieved for the layer between Cr(VI) ions and adsorbents. The authors finally expressed that all studied adsorbents showed different behavior. The DLM with single monodentate inner-sphere complexation gave reasonable prediction for adsorption on ferrihydrite, while the combined DLM and TLM pointed out monodentate inner-sphere and outer-sphere complexations for Cr(VI) adsorption on kaolinite. The authors also declared that the partly Cr(VI) reduction simultaneously occured in adsorption process [22]. In another study, humic acid and NaCl effects on Cd(II) mobilization throughout soil layer in Cd-contaminated soil were investigated by Ondrasek *et al.* It was observed that the interaction between humic acid and Cd(II) resulted in decrease of retained Cd(II) amount; contrary, the presence of NaCl increased the phyto-uptake of Cd(II). The speciation of Cd(II) and organo-complexation in soil solution were performed by the aid of Visual MINTEQ program and NICA (Non-Ideal Competitive Adsorption)–Donnan model, respectively. The input parameters were total metal and dissolved organic carbon (DOC) concentrations, and pH of soil solution. It was explained that the humification degree of Cd(II) contaminated soil caused reducing bioavailability and increasing accumulation of toxic Cd(II). Consequently, the authors expressed that chloride ions and deprotonated humates complexed with bioavailable Cd(II) species, thus determined Cd mobility throughout soil matrix [23].

Holmström *et al.* were evaluated two chemical equilibrium models for determination of Al(III) speciation in soil solution throughout forest soil. These models are based on the interaction of Al(III) with organic acids. It was assumed that the Al(III) binds to both low molecular weight organic acids and high molecular weights acids in soil solution. The results calculated from models were compared with the experimentally measured free Al ion and Al amount bound to low molecular weight organic acids. The Windermere Humic Aqueous Model (WHAM) is a combined model of Humic Ion-Binding Model V (defined as a discrete site/electrostatic model for H^+ and metal ion binding to humic and fulvic acids) and models including inorganic salts, aluminum, and hydrated iron oxide precipitations.

The findings of many studies, purposing the determination of equilibrium constants of proton-organic matter and metal-organic matter interactions, involve in WHAM database. All equilibrium constants of complexation between low molecular weight organic acids and inorganic salts (i.e., phosphate and sulfate) that exist in this database due to the same complex reactions are valid. In another model which is a simple chemical equilibrium model performing by the aid of MINEQL program, humic acids and fulvic acids are considered as monoprotic weak acids. In running of MINEQL, the charge ratio of inorganic ions such as chloride and sodium ions was considered as non-reacting species, while the pH of solution, concentrations of organic acids, and the total concentrations of Fe, Al, Mg, and Ca were used as input values. It was assumed that the total Al corresponded to sum of free Al(III) and Al(III) bound with high and low molecular weight organic substances, and Al(III) compounds with hydroxyl, phosphate, and sulphate. As a result, it was concluded that the Al in soil solution of mor layer mostly complexed with fulvic substance and low molecular weight organic substances [24].

Pd, Rh, and Pt sorptions on semi-arid soil and sediment samples were investigated and modeled by Sako *et al.* (in 2009). The acidity constants determined from titration data were approximately 5 and −7 for $LogKa_1$ and $LogKa_2$, respectively. The conditional binding constants of metals to surface were calculated by the aid of FITEQL program. The highest constant value was obtained for Pd whom the surface has high affinity. It was implied that the highest adsorption affinity was observed for the sediment sample having highest clay content and exchangeable cations density. According to authors, adsorption at low pHs could be attributed to interaction of metal with permanent negatively charged sites, and also, the conformity between experimental and theoretical values showed the validity of established model. The acidity constants of surfaces were optimized by nonlinear optimization equilibrium program using DLM. On the other hand, FITEQL4.0 was used to optimize the binding constants inputing acidity constants, pH values, total concentrations of metal, and active sites. The concentration of active sites was calculated from cation exchange capacity and concentrations of solids. In this study, MINEQL+ program was also used for optimization of Pd, Rh, and Pt adsorptions on sorbents. The acidity constants, concentrations of active sites, specific surface areas of adsorbents, and DLM were the input values. Both modeling studies gave reasonable results, conforming to the experimental results. As a result of the study, the authors declared that their SCM approximated to real complex retention process of platin group elements in soil with the low goodness of fit values [25].

Cd(II) adsorptions onto goethite-*Pseudomonas putida* composites having different mass ratio were modeled by the aid of FITEQL 4.0 using component additivity (CA) approach [26]. The theoretical results was coherent with the experimental results for 5:1 (goethite: *P. putida*) composite, while those were 8% over for 1:1 goethite:*P. putida* composite at high Cd(II) concentrations. This discrepancy was removed by adding site blockage reactions between *P. putida* and goethite, named as "CA-site masking" models [26]. As a result, it was mentioned that the CA method could be applied for modeling Cd(II) adsorption on iron oxides–bacteria composites provided that the concentrations of bacteria and Cd(II) were low, and when the concentration were high, interfacial complexation reactions between mineral and bacteria should be considered. On the other hand, Ca(II), Pb(II), Cd(II), and Cu(II) adsorptions onto iron and aluminum hydroxides were simulated by using Charge Distribution Multi-Site Complexation (CD-MUSIC) model. The authors described the non-specific adsorption on the basal surfaces as bidentate complexation and the specific adsorption on edge sites as monodentate inner-sphere complexation. The affinities of surface for protons and electrolyte ions, and site densities were used as input values in CD-MUSIC model assuming singly and triply coordination with surface oxygens [26]. Similarly, Pb(II) adsorption onto montmorillonite-bacteria composite having different mass ratio by combining surface characterization method [such as EXAFS (X-ray absorption fine structure spectroscopy) and ITC (isothermal titration calorimetry)] and surface complexation model by Qu *et al.* (2018). As a result, it was determined that the when the concentration of Pb(II) ions was low, it binded to phosphoryl groups of bacteria, while it binded to carboxyl groups at high initial concentrations. Thus, Pb(II) ions acts as a bridge by forming surface-metal-bacteria ternary complex [27].

Pb(II) mobility in soil system was investigated simulating adsorption on low-crystalline aluminum silicate (LCAS) in terms of adsorption effectiveness and surface complexation depending pH, ionic strength, and initial concentration of Pb(II) by considering extended TLM (ETLM). Pb(II) adsorptions on goethite, hydrous ferric oxide (HFO), quartz, and LCAS were assumed to be occurred by formation of inner-sphere monodentate or bidentate ternary complexes with anions of electrolyte solution, and it was decided that the inner-sphere bidentate complex is dominant above neutral pHs. On the other hand, at lower pHs, the inner-sphere monodentate complex is dominant with high surface coverage. Surface acidity properties of sorbent, determined by acidimetric and alkalimetric titration of sorbent in the presence of inert electrolyte, is primary required parameter in application of ETLM. The further input parameters are surface area, site density, adsorption constants of electrolyte, and capacitance of inner and outer layers. The adsorption constant, which was calculated from model, conformed with those of goethite; thus, it was envisaged by the authors that the Pb(II) adsorption on LCAS resembled to adsorption on goethite, and HFO, LCAS, and goethite were efficient adsorbents for Pb(II) uptake or immobilization [28].

Gavrić *et al.* modeled the retention of Cu(II), Pb(II), and Zn(II) on soil obtained from three different depth of grass swales by using source-based urban runoff model (StormTac Web) allowing long-term calculations with few input data. Thus, the simulating of annual heavy metal pollutant mass retained from swale soils was carried out. The rain and snow precipitation, the land-use data (such as corresponding runoff contributing area and the volumetric runoff coefficient), and the StormTac Web database were inputing to program. Then, the calculated metal uptake (MBs) was compared with measured one (MBm), and the results were given as MBs/MBm. In order to calculate the total loaded heavy metal on swale, the annual loads yielded from model were multiplied by the age of swales. As a result, the authors declared that the model needed soil chemistry data for studying rehabilitation of older swales [29].

On the other hand, these kinds of models have been used for determining kinetic properties of adsorption and/or desorption. For this purpose, Peng *et al.* (in 2018) used WHAM 7 and CD-MUSIC models to define kinetics of Cd, Cu, Ni, Pb, and Zn adsorption on soil. Humic acid and fulvic acid were specified as primary components of soil organic material (SOM). WHAM 7 was used for describing the reactions between heavy metals and SOM. Carboxylic and phenolic functional groups of humic and fulvic acids were assumed as active sites which the metal ions formed mono-, bi-, and tri-dentate complexes with. Totally, eleven complexations were estimated with distinct formation constants. Metal ions may give further reactions with hydrated aluminum and iron oxides to form monodentate complexes. CD-MUSIC model was especially used for explaining the interaction between heavy metal ion and hydrated iron oxides. The kinetic parameters for adsorptions and desorptions were calculated for all specified reactions by the aid of models. Consequently, rather different rate constants of adsorptions and desorptions showed that those individual sites became activated at different time scales. It was expressed by the authors that this situation determined the fate and bioavailability of heavy metals in soil environment [30].

The molecular dynamics (MD) computer simulations were used to quantify the mechanisms of adsorption and desorption of ions on illite, smectite, and interstratified illite/smectite (I/S) clays. The adsorptions of metal cations on priorly identified several structurally different adsorption sites were investigated in terms of adsorption free energy, preferable sorption sites, and their surface distributions. The authors expressed that the calculated equilibriums conformed with the literature knowledges [31].

9.3 Modeling of Sorption of Other Pollutants onto Soil or Soil Components

The competition between fulvic acid and arsenate adsorption onto active sites of soil was investigated by using 1-pK Three Plane Model with the CD-MUSIC option (TPCD) which includes two Stern layers at diffuse layer and surface interface, each Stern layers have their own capacitance. The model was applied by Visual MINTEQ. It was assumed that the fulvic acid binded to surface by forming monodentate complex with hydrated oxide surface via carboxylic acid functional groups. The optimization of carboxylic groups was performed by using K_d value of arsenate. Visual MINTEQ runnings showed that the predicted values are quite close to experimental values. As a result it was declared that the presence of humic substance decreased the arsenate adsorption meaning humic substance was a primary competitor for anion adsorption on soils [32].

It has been known that the persistent organic pollutants (POPs) react with soil organic matter (SOM). Ahmed *et al.* (in 2014) inspected hexachlorobenzene (HCB) adsorption on soil. They developed a model which separated representative system including functional groups and fractions of SOM. Additionally, Py-FIMS and XANES molecular analyses were performed due to HCB interactions with SOM via functional groups or molecular components of SOM. Phenol, catechol, 3,4,5-trimethoxy cinnamic acid, benzene, ethylbenzene naphthalene, ethylnaphthalene, glucose, glycine, hexaglycine, ethylnitrile, pyrrole, pyridine, acetic acid, alkane, alkene, acetamide, acetaldehyde, dimethylketone, methylacetate, methylamine, and aniline were chosen as representative SOM compound and/or functional groups in model. Then, the preferential interactions between HCB and components were determined considering their geometries. The HCB-SOM components interactions evaluated by quantum chemical density functional theory. The quantitative structure–activity relationship (QSAR) was used to correlate the bonding energies of HCB-SOM complexes to various physical properties of representative SOM components. As a result of QSAR analysis, it was understood that the physical properties of SOM such as polarizability, molar volume and mass, charge, and percentage of carbon atoms were primary parameters designated the interactions between HCB and SOM components [33].

Similarly, thymol adsorptions on montmorillonite and kaolinite were modeled by theoretical finite multilayer with multisite occupancy models. According to authors, HYBRID error function, providing the lowest sum of normalized error for all parameters, proved that the model described thymol adsorption better than Langmuir and Freundlich models. The model assumed that there were layers on the surface which the adsorbate retained on, and each layer had own adsorption energy level. The layer may be empty, or one layer or more may be occupied. When the number of layer is more than one, multimolecular adsorption occurs with a vertical position. On the other hand, if the layer number is less than one, multianchorage adsorption take places with a parallel position. As a result of modeling studies, two layers were determined for Fe(III) modified montmorillonite [34].

Tetracycline adsorption on soils was inspected by multiple linear regression modeling by considering soil properties such as organic carbon content and cation exchange capacity. While establishing the model, soil characteristics was used as input data. The predictions obtained from model gave satisfactory results. Consequently, the authors declared that tetracycline strongly retained on soil having high organic matter and acidic character, while its mobilization enhanced in alkaline soils which had low organic matter and clay content [35].

9.4 Conclusion

The modeling of sorption of pollutants as heavy metal and organics has great importance as regards of estimation their behavior throughout soil layers and their running into groundwater. The results of modeling studies simplify the understanding the state of clays and soil. Further advantage of modeling studies is the capable of clarifying the mechanisms of pollutant mobility, distribution, and bioavailability of pollutant in the natural environment. The surface complexation model, which mostly defines the interaction between heavy metals and active sites of soil, has been used by applying by the aid of PC programs such as FITEQL, MINEQL, and MINTEQ. Thus, the feasibility of quantitative evaluations of the interaction of pollutant and soil allows better understood of complex interactions of aqueous pollutant and soil or soil component. Consequently, in order to obtain higher quality databases about soil system and to clarify the condition of soil, the modeling studies should be enhanced by the authors.

References

1. Nortcliff, S., Hulpke, H., Bannick, C.G., Terytze, K., Knoop, G., Bredemeier, M., Schulte-Bisping, H., Definition, Function, and Utilization of Soil, in: *Ullmann's Encyclopedia of Industrial Chemistry*, pp. 399–420, Wiley-Vch Verlag Gmbh & Co. Kgaa, Weinheim, Germany, 2012.
2. Demirçivi, P., Karadirek, Ş., Anilanmert, B., Tutar, A., Hizal, J., Apak, R., Modelling of acidic and complexation properties of humic acids using the Irwing-Rossotti method. *Fresen. Environ. Bull.*, 24, 4743, 2015.
3. Karadirek, Ş., Kanmaz, N., Balta, Z., Demirçivi, P., Arda, A., Hizal, J., Apak, R., Determination of total antioxidant capacity of humic acids using CUPRAC, Folin-Ciocalteau, noble metal nanoparticle- and solid-liquid extraction-based methods. *Talanta*, 153, 120, 2016.
4. Boily, J.F. and Fein, J.B., Adsorption of Pb(II) and benzenecarboxylates onto corundum. *Chem. Geol.*, 148, 157, 1998.
5. Kosmulski, M., How to handle the ion adsorption data with variable solid-to-liquid ratios by means of FITEQL. *Colloids Surf.: Part A*, 149, 397, 1999.
6. Kriaa, A., Hamdi, N., Goncalves, M.A., Srasra, E., Acid-base Properties of Tunisian Glauconite in Aqueous Suspensions. *Int. J. Electrochem. Sci.*, 4, 535, 2009.
7. Zhang, Y., Jiang, J., Chen, M., MINTEQ modelling for evaluating the leaching behavior of heavy metalsin MSWI fly ash. *J. Environ. Sci.*, 20, 1398, 2008.
8. Jeske, A., Use of innovative and advanced computer simulations of chemical speciation of heavy metals in soils and other environmental samples. *Soil Sci. Annu.*, 65, 65, 2014.
9. Hızal, J. and Apak, R., Modelling of cadmium(II) adsorption on clay minerals individually and in the presence of humic acid. *Appl. Clay Sci.*, 32, 232, 2006.
10. Gu, X. and Evans, L.J., Surface complexation modelling of Cd(II), Cu(II), Ni(II), Pb(II) and Zn(II) adsorption onto kaolinite. *Geochim. Cosmochim. Acta*, 72, 267, 2008.
11. Kalantarı, K., Ahmad, M.B., Masoumi, H.R.F., Shameli, K., Basri, M., Khandanlou, R., Rapid and high capacity adsorption of heavy metals by Fe_3O_4/montmorillonite nanocomposite using response surface methodology: Preparation, characterization, optimization, equilibrium isotherms, and adsorption kinetics study. *J. Taiwan Inst. Chem. Eng.*, 49, 192, 2015.
12. Abdel Karim, A.M., Zaki, A.A., Elwan, W., El-Naggar, M.R., Gouda, M.M., Experimental and modelling investigations of cesium and strontium adsorption onto clay of radioactive waste disposal. *Appl. Clay Sci.*, 132, 391, 2016.
13. Siroux, B., Beaucaire, C., Tabarant, M., Benedetti, M.F., Reiller, P.E., Adsorption of strontium and caesium onto an Na-MX80 bentonite: Experiments and building of a coherent thermodynamic modelling. *Appl. Geochem.*, 87, 167, 2017.

14. Mockovčiaková, A., Orolínová, Z., Škvarla, J., Enhancement of the bentonite sorption properties. *J. Hazard. Mater.*, 180, 274, 2010.
15. Bickmore, B.R., Rosso, K.M., Nagy, K.L., Cygan, R.T., Tadanier, C.J., AB initio determination of edge surface structures for Dioctahedral 2:1 phyllosilicates: implications for acid-base reactivity. *Clays Clay Miner.*, 51, 359, 2003.
16. Hızal, J. and Apak, R., Modelling of copper(II) and lead(II) adsorption on kaolinite-based clay minerals individually and in the presence of humic acid. *J. Colloid Interface Sci.*, 295, 1, 2006.
17. Schaller, M.S., Koretsky, C.M., Lund, T.J., Landry, C.J., Surface complexation modelling of Cd(II) adsorption on mixtures of hydrous ferric oxide, quartz and kaolinite. *J. Colloid Interface Sci.*, 339, 302, 2009.
18. Arda, D., Hizal, J., Apak, R., Surface complexation modelling of uranyl adsorption onto kaolinite based clay minerals using FITEQL 3.2. *Radiochim. Acta*, 94, 835, 2006.
19. Hızal, J., Apak, R., Hoell, W., Modelling Competitive Adsorption of Copper(II), Lead(II) and Cadmium(II) by Kaolinite-Based Clay Mineral/Humic Acid System. *Environ. Prog. Sustainable Energy*, 28, 493, 2009.
20. Hizal, J., Demirçivi, P., Karadirek, Ş., Apak, R., Investigation of Individual and Competitive Adsorption of Cu(II), Cd(II) and Pb(II) on Montmorillonite in terms of Surface Complexation and Kinetic Properties of Cu(II) Adsorption. *Desalin. Water Treat.*, 57, 22441, 2016.
21. Vithanage, M., Rajapaksha, A.U., Dou, X., Bolan, N.S., Yang, J.E., Ok, Y.S., Surface complexation modelling and spectroscopic evidence of antimony adsorption on iron-oxide-rich red earth soils. *J. Colloid Interface Sci.*, 406, 217, 2013.
22. Veselská, V., Fajgar, R., Číhalová, S., Bolanz, R.M., Göttlicher, J., Steininger, R., Siddique, J.A., Komárek, M., Chromate adsorption on selected soil minerals: Surface complexation modelling coupled with spectroscopic investigation. *J. Hazard. Mater.*, 318, 433, 2016.
23. Ondrasek, G., Romic, D., Rengel, Z., Interactions of humates and chlorides with cadmium drive soil cadmium chemistry and uptake by radish cultivars. *Sci. Total Environ.*, 702, 134887, 2020.
24. Holmström, S.J.M., van Hees, P.A.W., Lundstrfm, U.S., Modelling of aluminium chemistry in soil solution of untreated and dolomite treated podzolic soil. *Geoderma*, 127, 280, 2005.
25. Sako, A., Lopes, L., Roychoudhury, A.N., Adsorption and surface complexation modelling of palladium, rhodium and platinum in surficial semi-arid soils and sediment. *Appl. Geochem.*, 24, 86, 2009.
26. Qu, C., Ma, M., Chen, W., Cai, P., Yu, X.Y., Feng, X., Huang, Q., Modelling of Cd adsorption to goethite-bacteria composites. *Chemosphere*, 193, 943, 2018.
27. Qu, C., Du, H., Ma, M., Chen, W., Cai, P., Huang, Q., Pb sorption on montmorillonite-bacteria composites: A combination study by XAFS, ITC and SCM. *Chemosphere*, 200, 427, 2018.

28. Usiyama, T. and Fukushi, K., Predictive model for Pb(II) adsorption on soil minerals (oxides and low-crystalline aluminum silicate) consistent with spectroscopic evidence. *Geochim. Cosmochim. Acta*, 190, 134, 2016.
29. Gavrić, S., Larm, T., Österlund, H., Marsalek, J., Wahlsten, A., Viklander, M., Measurement and conceptual modelling of retention of metals (Cu, Pb, Zn) in soils of three grass swales. *J. Hydrol.*, 574, 1053, 2019.
30. Peng, L., Liu, P., Feng, X., Wang, Z., Cheng, T., Liang, Y., Lin, Z., Shi, Z., Kinetics of heavy metal adsorption and desorption in soil: Developing a unified model based on chemical speciation. *Geochim. Cosmochim. Acta*, 224, 282, 2018.
31. Kalinicheva, A.G., Loganathan, N., Wakou, B.F.N., Chen, Z., Interaction of ions with hydrated clay surfaces: Computational molecular modelling for nuclear waste disposal applications. *Procedia Earth Planet. Sci.*, 17, 566, 2017.
32. Gustafsson, J.P., Arsenate adsorption to soils: Modelling the competition from humic substances. *Geoderma*, 136, 320, 2006.
33. Ahmed, A.A., Kühna, O., Aziz, S.G., Hilal, R.H., Leinweber, P., How soil organic matter composition controls hexachlorobenzene- soil-interactions: Adsorption isotherms and quantum chemical modelling. *Sci. Total Environ.*, 476, 98, 2014.
34. Nakhli, A., Mbouga, M.G.N., Bergaoui, M., Khalfaoui, M., Cretin, M., Huguet, P., Non-linear analysis in estimating model parameters for thymol adsorption onto hydroxyiron-clays. *J. Mol. Liq.*, 244, 201, 2017.
35. Conde-Cid, M., Fernández-Calviño, D., Nóvoa-Muñoz, J.C., Núñez-Delgado, A., Fernández-Sanjurjo, M.J., Arias-Estévez, M., Álvarez-Rodríguez, E., Experimental data and model prediction of tetracycline adsorption and desorption in agricultural soils. *Environ. Res.*, 177, 108607, 2019.

10

Soil Chemistry: Composition and Laws

Manju Yadav[1] and Dinesh Kumar[2*]

[1]*Department of Chemistry, Dronacharya Govt. College, Gurugram, India*
[2]*School of Chemical Sciences, Central University of Gujarat, Gandhinagar, India*

Abstract
Gain knowledge about the mysteries of soil chemistry and its function in farming and the atmosphere. Investigate the basic laws of soil chemistry, whether they impact absorption, transfer of cations and anions, and other processes. Explore how the most powerful powers in adsorption, chelation, etc., which involve influences of implementation speed, reaction time, weather effects, pH-to-salt saturation associations, and desorption. This chapter covers the accumulation of electric charge in the soil, which generates electrical and chemical potentials so that ions pass through barriers such as root membranes and plant bodies.

Keywords: Heterogeneity, organic matter, soil goethite, micronutrients, macronutrients, mineral surfaces, accumulation, unsaturated media

10.1 Introduction

"No single respects what is at his bases; we all stare at the stars." [Quintus Ennius (239–169 BC)]. "Heaven is beneath our feet and above our heads." [Henry David Thoreau (1817–1862)]. These quotations illustrate how differently humans see the soil that gives them life and feeds them. People's feelings have been whispered for an extended period. Maximum people are quiet at the awareness level of Quintus Ennius, who survived over 2,000 years back. They take for granted the diet that the soil yields, the hygienic water, and air that the soil delivers. Until now, the soil is amazing if one gazes sensibly. The permeable surface of the earth to nearby 1-m deep, the deepness that roots pierce, has much anonymity. The soil is as

**Corresponding author*: dinesh.kumar@cug.ac.in

Inamuddin, Mohd Imran Ahamed, Rajender Boddula and Tariq Altalhi (eds.) Applied Soil Chemistry, (183–196) © 2021 Scrivener Publishing LLC

cryptic and thrilling as any further science and any other portion of the earth. The soil heterogeneously combines inorganic and natural solids, water, air, microorganisms (both animals and plants), and plant roots. All these phases influence each other: soil chemistry treated these processes of chemical reactions. That is to say, carbon dioxide in the air associated with water acts against inorganic solid phase weather [1, 2]. Both plant growth and water quality are impacted by chemical processes between soil solid and soil solution. Bacteria stimulate many reactions, and plant roots grip and exude inorganic and organic materials. Soil chemistry has conventionally focused on chemical processes in the soil to effects the growth of plants and plant nutrition. Though established in the 1990s as anxieties rise around an inorganic and inorganic impurity in water and soil and their influence wildlife crop and human health, soil chemistry is now tended to focus on soil chemistry in the environment.

Quantum mechanics, the science of atomic and subatomic systems, are underlying principles of many fields in physics and chemistry. It is required for understanding the behavior of the systems. It established this field of science during the beginning of the 20th century by several scientists, including Neil Bohr, Max Planck, and Albert Einstein. The basic of chemistry is concerned with studying atoms, elements, molecules, and their transformation into matter or substances, ions, and salts. Chemical bonding, energy transmission, and reactions are tangled in generating these conversions. All these become the established science after the discovery of the law of conservation of mass by Lavoisier. This law shows that chemical reactions are balanced reactions and occur in an equivalent amount. This concept of the atom is the smallest particle of matter was first postulated by Democritus, Greek philosophers in 425 BC approximately. John Dalton was generally credited as a founder of atomic theory. Since then, crooks, Thomson's in 1897, working on the conduction of electricity in rarefied gases, made a revision of theories and conclude that it composed the atom of still smaller particles.

10.2 The Materials With Which Nutrients React

Apart from organic matter, it is commonly acknowledged that nutrients react by clay minerals and metal oxides. This vision is very generalized, if just since each of these groups interacts with the others in such a way that pure instances of them may be uncommon in soil. Readily available many swot up of the nutrient effect along with each one of these content groups, and they facilitate to explain chemical response associated, yet they often created misconceptions due to a change between pure compounds and soil compounds.

10.2.1 The Clay Minerals

Clay mineral was the "original" resource in soil chemistry since their arrangement was elaborated near the beginning in the science's development, and likewise, their involvement in transferring ions was also realized quite quickly. Perhaps, the greatest-understood feature of soil science is the involvement of ion transfer in processes of K+, Na+, Ca2+, and Mg2+ with clay miners [3–5]. Studying ion exchange as a meaningful method for the reaction of metallic cations like copper and zinc thru the soils was normal. The accumulation in soil solutions of these micronutrients metal ions is ample, lesser than that of the chief cations and particularly poorer than calcium. Because of the extreme role of valence in transferring ions, ions here within soil solution, when examining exchange site rivalry, then calcium is found to be the utmost value. Ion interchange could only be a significant process for micronutrient metals if there was an excess of options for copper as for calcium. This does not appear to be the instance [6]. This creates it improbable that transferring ions is a necessary process for metals for micronutrients. Other sinks are "deeper" for these ions, and there are a few ions left to play a part in the ion exchange.

Though, it frequently makes efforts toward evaluation of transferable metal cations. Commonly, this requires a preliminary pulling out of the soil by water to take away soluble metal. However, the key phase is to extract the "transferable" metal with a salt solution. Understanding the outcomes found to be argumentative with two motives. The utmost recognizable the first one is worried through the pH consequence. If the pH is not regulated, the salt solution will reduce the pH for the most soils and will also produce several sorbed metal since of the influence of pH on sorption. If the pH is controlled at 7 [7], this will alter the pH at the majority soils, and this shift will influence the sorption power or, thus, the noticeable value of the transferable ion. To order to explain the next motive, several facts lying on the impact of salt on covalent sorption need to get ahead. The universal law aims to increase the growing amount of electrical potential near a charged layer by raising the salt concentration. This turns out to be less damaging if the opportunity is deleterious, like that of the surface sorption of metal ions. This reduces metal sorption. Sources include kaolinite nickel sorption [8], soil zinc [9], copper, cadmium, and kaolinite lead [10], and for cadmium on soil [11]. Therefore, enhancing the salt concentration to eliminate evidently transferable ions will, therefore, eliminate selected metal ion that is adsorbed covalently. This is a situation where specific materials may even have created an issue. Most covalent sorption education programs have included oxides of iron. Research lab gets ready oxides which have a limit of null charges > 8, the absolute importance depending very much on the degree of the imperfection among carbonates. Metal sorption normally occurs on a pH greater than eight and hence lying on the productive face. The individual consequence of growing ionic potency is toward the improvement of sorption according to the stated law. It's that ionic strength also reduces the coefficient of action, however, and this aspect reduces sorption so that the overall impact is minimal. Thus, no influence of electrolyte accumulation on the adsorption by iron or aluminum oxides of copper and zinc were identified [12]. Similarly, they inferred from their own research and from a literature survey [13] that the impact of concentration of electrolyte taking place on metal adsorption remained marginal. Its solitary case where data gathered from pure compounds can never be applied to the soil clearly.

10.2.2 Variable-Charge Mineral Surfaces

The clay minerals are rare crystals, considering their identification with soil experts. It is because the interactions of face atoms are perfectly delighted throughout their surface area. For most crystals, that's not the case. The "modern" example is silver chloride that mostly consumes the overindulgence of silver or chlorine ion, focusing on whether medium produces additional silver or else further chlorine. The constituent part, therefore, holds a charge dependent on the overall structure of the medium. Metal oxides, including iron, copper, and manganese, still have unmet associations to the exterior of every crystal. Surfaces of soil particles like dissimilar on or after the sides of the planes. Such surfaces take up molecules of water to break such bonds. Such water particle which put on or drops hydrogen ion (H^+), moreover, thus gain a charge and focusing on the pH. Thus, they are adjustable charge surfaces with load based on the pH. Several oxides are found in soils. Researchers listed six-iron oxides, five aluminum oxides, and its 13 oxides of manganese. (The word "oxide" is being used broadly here now to describe oxyhydroxides.) Goethite [14, 15] of iron oxides is quite usually found in soils in several atmospheric areas. The survival of oxides would only be evident from either the detection of one of the most evident father properties of soils: their color. The color of soil is partially associated with the quantities of the gradual substance in the soils, however, as well toward the quantity and unit extent of the oxides existing. Oxides of iron are only significant by goethite providing haematite the reds along with the yellow-browns. The terms of red land, terra rosse in addition to kraznozem that already attributed to soils in verbal communication and dialects, the entire show the potency of reddening within attracting the consideration of cultures. At hand, it comprises a limitless study of ions sorption by different charge minerals, as well as micronutrients, macronutrients, and contaminants. Scientists listed >50 inorganic cation and anion reaction experiments with hydrous ferric oxides. Anions of Mg, Hg, Ag, Co, Ni, Cd, Mn, Pb, Ca, and Sr; the cations of PO_4, AsO_4, SO_4, SeO_3, SeO_4, VO_4, CrO_4, and $B(OH)_3$ are examined. These records compose it understandable so as to this effect might be significant in dominant the contribution of necessary resources and poisonous particles to plant equally. Goethite is the most widely researched iron oxides in the rock. We allow many properties of goethite to be clarified to recognize the inferences and constraints of those findings. It is known as α-FeOOH, where α corresponds to closing hexagonal packaging. The goethite in the soil is, however, dissimilar from the substance used in several lab experiments. Far from being unadulterated, it may include manganese and other trace metals [16, 17], different anions [18], and, in particular, aluminum [19–23] defined as rare fibrous goethite with atomic ratios of Fe:Al:Si is 4:2:1. It leads to a weight of about 5% Si [24]. The silicon content in iron oxide pellets was derived from a range of soils varying from 1% to 4%. It's hard to be sure that those measurements are perfect, but those values should be clear of this issue. To be conscious of soil goethite behavior, we need to learn how all these silicon atoms are organized in an appropriate manner. It just seems unlikely they'll be dispersed at casual through the crystal; however, this is hard to examine for just the tiny soil goethite particles [23]. The silicon, actually, is use in a stable place since this was opposed to oxalate removal. Researchers hypothesized to perhaps the adsorption of silicon onto the microcrystal exterior. This recommendation is compatible with the company of research moreover [25], who examined four tests of botryoidally goethite ranging from nil to 1.25% in silica material. The silica-containing samples were made from perfect syringe-like fiber to widths as minuscule as 30 μm. We recommended the gap among the grains may comprise a silica monolayer. The structure was not present in the goethite sample; they did not involve silica. The silica in soil goethites may live among the oxide micro-crystals as a variety of mortar. Because these pollutants occur in ordinary goethites, whose activities can vary from those of ordinary goethite, often observed. The presence of silica, for example, implies that the shift in charge and ability through modification of pH would be distinct from those of pure goethites. Goethite's water content is normally better compared to measure as of the FeOOH calculation. It is accurate of artificial goethite [26] and of natural goethites [18]. One justification seems to be that, as suggested by several researchers [27], the additional water may be an H_2O particle encircled into the arrangement. Here, the added liquid might be linked between field edges in micro-pores. Perhaps, the theory is that the system requires certain Fe atoms, and charge offset via superfluous protonation of neighboring oxygen part [28].

10.2.3 Organic Matter

Reactions of organic substances with copper and zinc comprise repeatedly considered. One motivation meant in favor of these learning is the general happening of deficiency in crude soil. Of either the copper has the highest preference used for organic material, along with even several research has established as organically associated equally to half of the copper current. It is complicated to create some uncomplicated announcement regarding the series of inclination meant for metals. A standard series of affinities is found to be Mg < Mn < Zn < Ca < Co < Pb < Ni < Cu. Some point, however, if such combinations are contradictory and rely on the organic matter environment, the tools being used, and pH measurements. A significant issue is the level of freight. Therefore, Cd is powerfully favored toward calcium by slight loadings; however, its resemblance is comparable to high loadings. For the conclusion that whether research teams analyze outer-sphere complexes, chelates, and the inner-sphere complexes be likely to depend on the laboratory circumstances, it is difficult to make strong declarations as to the bond associated. He argues that Cu^{2+} mainly forms complexes within the inner sphere, while most of the various transition metals in the first row (Mn^{2+}, Fe^{2+}, and Co^{2+}) form complexes outside the sphere. For a complete examination of erstwhile metal bonding, [29] an essential feature of the interaction linking metal ions or organic earth soil matter is with the intention of the obvious similarity declines like the number of metal rises. The response is as if there are the minority sites to some extent, additional sites of rather an inferior affinity sites of modest affinity to some extent, and so on. Several strong affinity sites may include sulfhydryl assembly into organic soil matter. The outcome of the spot allocation is, when defining the chemical reaction, it may exist essentially to apply a word of affinity that diminishes as the number of loads creases, instead of a static affinity [30, 31]. The reactions between copper and fulvic acids [32] in the description of mercury contact with soil. In addition, it is important to note that such action presents a challenge of "chicken and egg". It stands for that single wants to be familiar by the quantity of loading previous to single can compute the mean affinity constant. Up till now, the individual desires the resemblance invariable to estimate the load. This trouble can be fixed by making several assumptions by saying, the affinity constant, determining the same load importance will provide moreover afterward testing whether the expected affinity constant will be prearranged by this loading. Computer systems can loop rapidly during a series of replicated calculations until the original estimated value coincides by the measured rate; this is denial extended trouble. Readily available seems to be a single aim to be drawn when allowing the nutrients to communicate with organic matter. Organic material is besides a changing charge matter: the charge changes by pH. They might assume that the various functional classes have different standards used for their pK, and they detach at a diverse pH rate. Here broad, the result be to the layer is constructive at mildly acidic and destructive at elevated pH. Thus, to explain soil reaction of "changeable load surface" is considered existing generic along with relevant to oxides, clay minerals plus organic matter.

10.3 Soil Chemistry: Laws

10.3.1 First Law

The laws of soil chemistry are shown in Scheme 10.1.

We are now wondering whether to apply it. Originally, we find the technique, which is now frequently employed, to be dismissed. Visualize a straightforward adsorption process in which a process agent is adsorbed on the surface of the particles, x. The equation of balance may then be interpreted:

$$\text{Site} + \text{a} \leftrightarrow \text{Site-a} \tag{10.1}$$

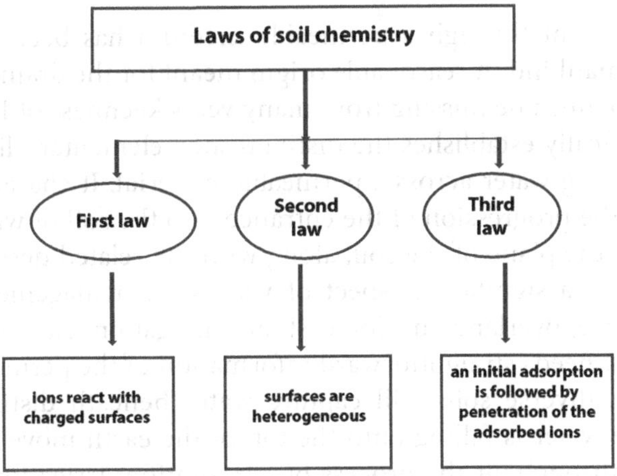

Scheme 10.1 The primary law—ions are reacting to activated surfaces.

If so, preliminary conc. of the site be p.

Stability concentration of a is x.

The product Site-a is E; afterward, the invariable K is known by this equation:

$$K = S/(p - E)\ x \tag{10.2}$$

Rearrange of above equation, we get

$$S = pKx/1 + Kx \tag{10.3}$$

It's the Langmuir equation of adsorption. This can be interpreted in many ways [33], although this basic derivation shows, it is consistent through adsorption lying on an identical common base, where every other site is identical and where the reaction does not modify. Thus, we will see it subsequently failures for soil nutrient reactions since it does not fulfill any of these needs. We'll believe for a moment if we should change this to include the primary law. It has developed a lot of models to illustrate the effect of positively and negatively charged ions, both by inconsistent charge oxides like goethite. The designs vary in convolution and objective. For instance, a few are involved with the complete modeling of the bonds developed on the surface. Though, all models comprise terminology so that the total reaction is not only dependent on the solution.

10.3.2 Second Law

Water movement through a permeable medium has been of immense concern to mankind. A reasonable origin meant for the drainage and irrigation mechanism be missing from many years keenness of Henry Darcy who scientifically establishes the distinguished elementary linear principle of passaging water across a permeable material. It characterizes permeation as the progression of the entrance into the soil of water through reachable at the plane of the soil, along with the related descending run. Permeation is a significant aspect of watershed management, groundwater recharge, overland run forecast, and irrigation. Resourceful water management needs straightforward information of the permeation speed upon which diverse soils will capture water beneath distinct circumstances. The water is falling onto the top of the earth move from unsaturated soil throughout the process of evaporation, irrigation, and plant extraction. Information on soil water run beneath unsaturated situation has highly developed speedily. In-depth, concentration should be now committed to the unsaturated intermittent liquid framework in a diverse porous medium. There are no constant flow patterns for some substantial period, so soil heterogeneity is the law instead of being an exception of existence in nature. Suppose that soils comprise separate layers of homogeneous topsoil. The pressure head and hydraulic head of the urbanized model; however, the moisture material is not supposed constant crossways the boundary of the two different layers. The strength of this method may be doubted since the differential equation of soil water runs is built beneath the supposition of how every dependency variable and its derivatives are nonstop. The equation of movement telling movement of water in soils is to get with substitute Darcy's rule in the continuity equation's differential form. However, just an integral continuity form. Scientists show a discontinuous alter the capillary pressure that can be present at the junction between dissimilar sand outside layer. We supposed that the pressure necessity is constant despite sudden amends in texture and incorrectly argues for a quick but permanent alteration of the pressure unconfirmed by pieces of evidence. Unless membrane pressure is not continuous, a steady pressure gradient will exist at the layer interface. Throughout this analysis, a process is accessible that illustrates soil heterogeneity by identifying that the soil's physical and hydraulic characteristics constantly clash at a reasonable depth. Many researchers used that same advance to the inconvenience of unsaturated flows that only saturated hydraulic conductivity was permissible to fluctuate by deepness.

Henry Darcy, in 1856, found the features of a downward run of water throughout saturated sand filters; Darcy's law states that, perhaps, the stream run via a column of saturated soil is directly proportional to only the loss of the head and inversely proportional to the duration of the flow direction. Darcy used specific gradients through saturated, homogeneous granular materials columns, and calculated the streams. The major limitations of this rule are that the rate of liquid should be relatively low to ignore friction forces; the surface and air must not interfere, and the liquid must be homogeneous and spherically symmetric.

Darcy's Law for Unsaturated Media: Although Darcy's method was established in the homogenous porous medium for saturation liquid, it has applicable to slightly saturated porous media issues. There were two significant factors acting on an item of soil water volume, which is the gravitational force forcing the item to collapse, and the capillary forces transferring the product from higher to lower pressure zones. Studies on the capillary movement of surface water found that water flow via the surface is like the rule of Fourier and electricity motion across a conductor, i.e., The Rule of Ohm. Up till now, we still observe much effort to match the equation of Langmuir to illustrate the definite soil ion sorption. While this equation needs a consistent reaction site, anyone uses it isn't only unwilling to detect the special outcome of exact sorption over ability but is still not in agreement with the proposal that surfaces of the soil are heterogeneous. A clear sign of variability derives from experiments that have demonstrated the existence of both positively and negatively charged in identical soil. The extent of the charge changes with pH, however, which can be there, most probably lying on distinct molecules or perhaps on the identical particle patch. It is an additional feature where the soil is at variance with oxides that are researched regularly as templates meant for soil. Oxides can be heterogeneous regarding Cation reaction but not anion reaction. This has been suggested [34] because certain absent Fe atoms are substituted inadequately by protons, providing locations of negative potential that would be observed with cations adsorbing but not with anions. Positively and negatively charged ions react through identical sites according to this criterion. That's never always the issue for the earth's soil. Positively charged ions should interact by its the most unconstructive ending of a potential continuum with anions have the most positive ending. Zinc and phosphate may react intensely with one another when each interacts to goethite, yet when interacting against soil, much of the impact of phosphate for zinc sorption is often not direct, and the reason is influences on pH. Specific effects are weak at weak zinc amounts since the two ions respond with opposite poles of the continuum of charged site. How we use the notion of heterogeneity is the issue to ask now. There were three separate methods used. Maybe the simplest thing to postulate is to make the surface of two or even more identical "under" surfaces. Similarly, the plane of hydrated ferric oxides was classified into superior and inferior-affinity exterior for reaction with cations, and this was celebrated as realistic, practical, and the Ockhamites.

10.3.3 Third Rule

The early reaction of absorption is pursued through the infiltration of adsorbed ions through the reactive particles within. It possibly provided the strongest support in model structures of this law. This demonstrated that Cd, Ni, and Zn sorption via goethite persisted meant for two days. Dissolution experiments, along with comprehensive simulations, also showed that it was compatible with an earlier adsorption reaction accompanied by a more recent work on diffusive surface penetration [35–37] showing with the aim of the mechanism used for this diffusion varies for various metals. Cd, Cu, Zn, and Mn appear to primarily pass across pores, appear to spread chiefly through pores. The existence of pores slightly affected nickel and chromium and will disperse through lesser defects such as gaps in crystals. All these mechanisms are relatively slow and have significant activation energy as they allow atoms to cross very high-power barriers. Hence, there is a broad temperature influence. The photo has just been clearer for anions, too [38, 39]. It has been shown that degree of the slow phosphate-reaction relied on the goethite crystalline. Specimens that were finely crystallized had a high reaction, but the reaction lasted for weeks with weakly crystallized samples. This provided clear proof that the process was measurable spatial diffusion across the crystal spheres. Inside this effort, diffusion brings to an end later than three weeks. That recommended to the piercing ions have arrived at the ends of their pathway. In certain situations, diffusion is consequently poor as compared with the scale of the responding material; the extent of the penetration is undersized. For this purpose, the penetrated sum should assume to be equal to the period's square root. This partnership can only be pursued unless the focus at the starting point is steady. As the transmission progresses, the quantity in solution reduces. As a result, the intensity of solubilized reactants declines, and because this is the remarkable concentration calculated for the diffusion phase. Many explanations exist into why not the experimental reaction obeys the square root rule. So, as we talk about the genuine recorded levels, those would be addressed. As charged unit communicates by the charged exterior, moreover, this is accompanied by diffuse penetration addicted to the material, a relevant issue is: Actually, what likely to take place to the charge so what's the subsequent consequences on the surface potential and its reply is, it changes surface electrical capacity. These necessities influence some fresh adsorption reaction and definitely several desorptions. This leads us to the next law, which may be spoken since: it is unfeasible to put fertilizer to the identical soil for two times. The soil is now transformed since it is impracticable. Clear support in favor of the next rule derives from macronutrient experiments as they are administered in adequate amounts to have measurable results. The encouragement comes from experiments where phosphate was once again applied to the surface, allowing the reaction to persist for an extended period as sorption scales were attaining on the reacting soil model, the curves evacuated obviously since the more phosphate present here. The curve slope also declined at definite concentrations, and this is the buffering capability.

Acknowledgments

Dinesh Kumar is thankful to DST, New Delhi, for financial support to this work (sanctioned vide project Sanction Order F. No. DST/TM/WTI/WIC/2K17/124(C).

References

1. Fierer, N., Bradford, M.A., Jackson, R.B., toward and ecological classification of soil bacteria. *Ecology*, 88, 1354–1364, 2007.
2. Fierer, N., Breitbart, M., Nulton, J., Salamon, P., Lozupone, C., Jones, R., Robeson, M., Edwards, R.A., Felts, B., Rayhawk, S., Knight, R., Rohwer, F., Jackson, R.B., Metagenomic and small-subunit rRNA analyses reveal the genetic diversity of bacteria, archaea, fungi, and viruses in soil. *Appl. Environ. Microbiol.*, 73, 7059–7066, 2007.
3. Jose, P.S.A., Patricia, F.L., Carlos, M.S., Modeling Sorbent Phase No ideality for the Accurate Prediction of Multicomponent Ion Exchange Equilibrium with the Homogeneous Mass Action Law. *J. Chem. Eng. Data*, 57, 6, 1766–1778, 2012.
4. Nathan, P., Graeme, J.M., Sara, J.C., de Bruyn, M., Kenneth, N., BDST modelling of sodium ion exchange column behaviour with strong acid cation resin in relation to coal seam water treatment. *J. Environ. Chem. Eng.*, 4, 2, 2216–2224, 2016.
5. Borba, C.E., Santos, G.H.F., Silva, E.A., Mathematical modeling of a ternary Cu–Zn–Na ion exchange system in a fixed-bed column using Amberlite IR 120. *Chem. Eng. J.*, 189, 49–56, 2012.
6. el-Sayed, M.H., Dzombak, D.A., Babcock, K.L., Thermodynamics of copper(II)- calcium exchange on bentonite clay. *Soil Sci. Soc. Am. J.*, 34, 397–400, 1970.
7. Shuman, L.M., Fractionation method for soil micronutrients. *Soil Sci.*, 140, 11–22, 1985.
8. Mattigod, S.V., Gibali, A.S., Page, A.L., Effect of ionic strength and ion pair formation on the adsorption of nickel by kaolinite. *Clays Clay Miner.*, 27, 411–16, 1979.
9. Barrow, N.J. and Ellis, A.S., Testing a mechanistic model. V. The points of zero salt effect for phosphate retention, for zinc retention and for acid/alkali titration of a soil. *J. Soil Sci.*, 37, 303–10, 1986.
10. Schindler, P.W., Leichti, P., Westall, J.C., Adsorption of copper, cadmium, and lead from aqueous solution to the kaolinite/water interface. *Neth. J. Agri. Sci.*, 35, 219–30, 1987.
11. Naidu, R., Bolan, N.S., Kookana, R.S., Tiller, K.G., Ionic strength and pH effects on the sorption of cadmium and the surface charge of soils. *Eur. J. Soil Sci.*, 45, 419–29, 1994.

12. Okazaki, M., Takomidoh, K., Yamane, I., Adsorption of heavy metal cations on hydrated oxides and oxides of iron and aluminium with different crystallinities. *Soil Sci. Plant Nutr.*, 32, 523–33, 1986.
13. Hayes, K.F. and Leckie, J.O., Modelling ionic strength effects on cation adsorption at hydrous oxide/solution interfaces. *J. Colloid Interface Sci.*, 115, 564–72, 1987.
14. Sposito, G., Derivation of freundlich equation for iron exchange reactions in the soil. *Soil Sci. Soc. Am. J.*, 44, 652–54, 1980.
15. Sposito, G., on the use of Langmuir equation in the interpretation of adsorption phenomenon. II. The two surface Langmuir equation. *Soil Sci. Soc. Am. J.*, 46, 1147–52, 1982.
16. Stiers, W. and Schwertmann, U., Manganese substitution in synthetic goethite. *Geochim. Cosmochim. Acta*, 49, 1909–11, 1986.
17. Bhattacharya, P., Hasan, M.A., Sracek, O., Smith, E., Ahmed, K.M., Von Bromssen, M., Huq, S.I., Naidu, R., Groundwater chemistry and arsenic mobilization in the holocene flood plains in south-central Bangladesh. *Environ. Geochem. Health*, 31, 1, 23–43, 2009.
18. Kuhnel, R.A., Roorda, H.J., Steensma, J.J., The crystallinity of minerals—a new variable in pedogenic processes. A study of goethite and associate silicates in laterites. *Clays Clay Miner.*, 23, 349–54, 1975.
19. Lewis, D.G. and Schwertmann, U., The effect of (OH) on the goethite produced from ferrihydrite under alkaline conditions. *J. Colloid Interface Sci.*, 78, 543–53, 1980.
20. Fey, M.V. and Dixon, J.B., Synthesis and properties of poorly crystalline hydrated aluminous goethites. *Clays Clay Miner.*, 29, 91–100, 1981.
21. Palani, K., Balasubramanian, K., Kalaivani, R.A., Study on aluminium contamination in mettur soil and its subsequent uptake by medicinal plants. *Orient. J. Chem.*, 34, 6, 3129–3133, 2018.
22. Schulze, D.G. and Schwertmann, U., The influence of aluminium on iron oxides, X. Properties of Al-substituted goethites. *Clay Miner.*, 19, 521–39, 1984.
23. Fordham, A.W., Merry, R.H., Norrish, K., Occurrence of microcrystalline goethite in an unusual fibrous form. *Geoderma*, 34, 135–48, 1984.
24. Fordham, A.W. and Norrish, K., Electron microprobe and electron microscope studies of soil clay particles. *Aust. J. Soil Res.*, 17, 283–306, 1979.
25. Smith, K.L. and Eggleton, R.A., Botryoidal goethite: a transmission electron microscope study. *Clays Clay Miner.*, 31, 392–6, 1983.
26. Thiel, R., Zum System a-FeOOH-a-AlOOH. *Z. Anorg. Allg. Chem.*, 326, 70–8, 1963.
27. Fey, M.V. and Dixon, J.B., Synthesis and properties of poorly crystalline hydrated aluminous goethites. *Clays Clay Miner.*, 29, 91–100, 1981.
28. Murad, E., Mossbauer spectra of goethite: evidence for structural imperfections. *Mineral. Mag.*, 43, 355–61, 1979.
29. McBride, M.B., Reactions controlling heavy metal solubility in soils, in: *Advances in Soil Science*, vol. 10, pp. 1–56, 1989.

30. Cheam, V., Chelation study of Cu(II) : fulvic acid system. *Can. J. Soil Sci.*, 53, 377–82, 1973.
31. Bresnahan, W.T., Grant, C.L., Weber, J.H., Stability constants for the complexation of copper (II) ions with water and soil fulvic acids measured by ion selective electrode. *Anal. Chem.*, 50, 1675–79, 1978.
32. Barrow, N.J. and Cox, V.C., The effects of pH and chloride concentration on mercury sorption. I. By a soil. *J. Soil Sci.*, 43, 305–12, 1992.
33. Elprince, A.M. and Sposito, G., Thermodynamic derivation of equations of the Langmuir type for ion equilibria in soils. *Soil Sci. Soc. Am. J.*, 45, 277–82, 1981.
34. Barrow, N.J., Brummer, G.W., Strauss, R., Effects of surface heterogeneity on ion adsorption by metal oxides and by soils. *Langmuir*, 9, 2606–2611, 1993.
35. Brümmer, G.W., Gerth, J., Tiller, K.G., Reaction kinetics of the adsorption and desorption of nickel, zinc and cadmium by goethite. I. Adsorption and diffusion of metals. *J. Soil Sci.*, 39, 37–52, 1988.
36. Barrow, N.J., Gerth, J., Brummer, G.W., Reaction kinetics of the adsorption and desorption of nickel, zinc and cadmium by goethite. II. Modeling the extent and rate of reaction. *J. Soil Sci.*, 40, 437–50, 1989.
37. Benkovitz, C.M., Scholtz, M.T., Pacyna, J., Tarrason, L., Dignon, J., Voldner, E.C., Spiro, P.A., Logan, J.A., Graedel, T.E., Global gridded inventories of anthropogenic emissions of sulfur and nitrogen. *J. Geophys. Res.*, 101, D22, 29239–29253, 1996.
38. Strauss, R., Brummer, G.W., Barrow, N.J., Effects of crystallinity of goethite: I. Preparation and properties of goethites of differing crystallinity. *Eur. J. Soil Sci.*, 48, 87–99, 1997.
39. Strauss, R., Brummer, G.W., Barrow, N.J., Effects of crystallinity of goethite: II. Rates of sorption and desorption of phosphate. *Eur. J. Soil Sci.*, 48, 101–114, 1997.

11

Parameters of Soil Chemistry

Manju Yadav[1] and Dinesh Kumar[2*]

[1]*Department of Chemistry, Dronacharya Govt. College, Gurugram, India*
[2]*School of Chemical Sciences, Central University of Gujarat, Gandhinagar, India*

Abstract
In today's world, the main three components of environment, *viz.*, soil, atmosphere, and water, be essentially a collection of ordinary wealth to support, develop, and enrich human life. Soil is, perhaps, the primary surface covering of the earth and is appropriate for plant production. Because of its importance for agriculture, the soil is a real concern. Water runs, land usage, and production of plants also have links to the soil. In the present chapter, discussion of a lot of essential soil ideas, and improvement, morphology, components, and belongings (soil air, water, colloids, ion exchange, i.e., cation exchange and anion exchange capacity and pH) will be made. Universal appreciative of earth ideas besides these interlocking associations is important toward creating a sound earth managing option.

Keywords: Parameter, component, colliods, aggregates, aeration, environmental impact, infiltration problems

11.1 Introduction

Soil is acting a very important character with wholesome existence resting on the earth. Soil like water impacts our lives in many methods and is important for the existence of all living systems. It behaves as a platform for crops that feed millions and millions of people every day that farming and grazing livestock. It is necessary to understand that soil and its characteristics influence existence, and even how soil has been changed and exploited by humankind. Soil awareness also strengthens our environmental understanding. Mismanagement of soil may lead to overall poor development

**Corresponding author*: dinesh.kumar@cug.ac.in

Inamuddin, Mohd Imran Ahamed, Rajender Boddula and Tariq Altalhi (eds.) Applied Soil Chemistry, (197–214) © 2021 Scrivener Publishing LLC

of any place. Environment plays a significant responsibility, directly or indirectly, both in strengthening and spoiling the soil. Soils are crucial for existence, in the sense that they supply the medium for plant development, home for a lot of insects and other organisms, proceed as a filtration method for surface water, carbon accumulates, besides safeguarding the environmental gas. They also hold up buildings and highways and contribute to the economies of our cities. Soils transport and move water, give homes for thousands of microorganisms and other creatures, and have lots of alternative arrangements of weathered rock and minerals. Still, no one modifies the chemistry of the soil quicker than humans execute. Several of these days' soil chemistry issues got to do with environmental sciences [1–5]. Soil chemist's analysis issues concerning organic and inorganic soil impurity, pesticides alternative pollutants, and environmental health threats. Soil value may be a major concern because of its significance to agriculture. Approximately, most of the food consumed by people, aside from obtained from aquatic ecosystems, is big inside the soils of the Earth. Many evident roles that soils offer to us, including paper and textile fiber, fuelwood manufacturing, and road or construction foundations. Less noticeable roles soils also act as a platform for microorganisms and biota to attenuate contaminants and waste nutrients, freshwater recycling, nitrogen recovery, and shelter. Soils often have many additional applications, including additives in pastries and cakes, insecticides, inks, brushes, maquillage, and medicines; application of clays varies from mud digging, ceramics, and works of art; to have shiny finishes on various paper items. Soil is also an essential part of almost any environment but is still overlooked. Soil should be called the base of the environment because soil fertility defines what an environment would look like regarding the animals and plants' existence it will support. The function that soil performs in forestry is also crucial to preserving water quality and site profitability over the long term. Soil conditions play an important role in crop production in agricultural fields. In urban regions, the soil has a dynamic role to play in lowering infiltration and nutritional uptake by overflow. The worth of the soil frequently becomes unnoticed before it tarnishes the soil, and it often reduces the essential soil resources are often until delivered. Soil chemistry involves the chemical processes and reactions within these elements and focuses, in particular, on researching the role of pollutants and nutrients in the soil. Awareness of soil science enables researchers to supervise, monitor. Knowledge of soil chemistry allows researchers to track, regulate, and predict the environmental impact of contaminants [6–9]. Chemical information joined with understandings from the Earth sciences, physics, and biology is desirable to recognize, avoid, and remediated environmental problems with soils.

11.1.2 Soil Morphology

A number of kinds express the soil morphology and configurations of the various layers that make up this and their visible and quantitative properties. The production of soil from the parental substrate resulting from processes of soil formation, which occurs under the influence of soil-forming factors. Such techniques can take time to produce soil, several thousand years. The soil develops and grows darker than time progresses and forms distinct structures called horizons. The horizon of the soil can be described as a horizontal layer, running parallel to the soil surface and with fairly homogeneous physical, biological, and chemical characteristics created by soil formation reactions. In the vertical section of the soil, several distinguishing horizons are able to be recognized. Such a segment is termed as the profile of the soil. It is generally considered from a level by 90° to superficial. The superior portion of a soil profile above the parent substance under which the cycle of soil-forming takes place and where the bulk of plant root and soil animals are located is named solum. Soil properties don't just shift vertically downward. Lateral differences occur, since the soil is a muti-dimensional structure, too. A three-dimensional soil model called a pedon is picked as a component to denote the existence and organization of horizons and the changeability of soil resources. A collection of identical pedons enclosed by non-soil or pedons with different identities in all directions is called polyhedron [10–12].

11.1.3 Basic Soil Components

It has five constituents, as seen in Figure 11.1. Mineral content comprises all minerals derived from parental substance and those designed in the soil solution through the rearrangement of materials. Organic matter is primarily produced from decomposing plant matter and decayed by the action of the soil's bacteria, animals, and microbes. It is this organic section that differentiates soil from geological matter occurring beneath the earth's plane, which or else may have many of the properties of soil. It usually fills the gaps in the soil are usually with air and water. Water and air in the soil have a collaboration relationship, although equally fighting for similar pore spaces. For instance, later than rain or if the soil is badly drained, the pores are packed with water, and the air is expelled. Alternatively, as H_2O travels

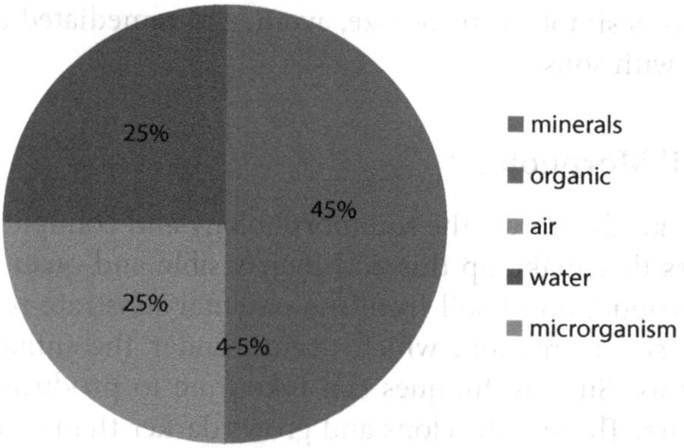

Figure 11.1 Soil composition by volume.

away from wet soil, the pore space is filled by air. Thus, the relationship between air and water in soils is constantly varying [13, 14].

11.1.3.1 Mineral

Mineral content comprises all minerals derived from either the parental substance, along with those produced by rearrangement of the soil solution from materials. The most significant soil factor is the mineral component which comprises more or less 45 to 49 of the quantity. Soil resources are components of two primary kinds of resources. Primary resources, such as those present in sands and limestone, unite the surface materials that unite the region from which they produced the same as the parent material. These are usually non-regular or oval. Counter minerals resulting from either the weathering of the first minerals and release the ions and shape several sustainable mineral forms such as salt clay. Clays have an outsized field which is very good for soil chemistry and the capacity to keep water. In adding together, negative and favorable charges near soil minerals influence the soil's capability to keep essential nutrients, like cations, leading to a soil's capacity to exchange ions (CEC).

11.1.3.2 Water

Water is the second basic constituent of soil. Water produces around 50% of the volume of the surface. Water is essential for transferring nutrients to plant and soil growth and for promoting biological and chemical decay. Accessibility of soil H2O is the capability of a soil to absorb water that is offered for plants. Soil's capacity to grab water is based on soil consistency.

The finer the particles in the soils, the more the soil repin with water. The clay soils have the highest water-holding potential and the lowest sands. The natural matter equally affects the water-holding potential of soils as the strong sensitivity of organic matter to water. The higher the amount of organic content in the soil, the greater the ability of soil to keep water. Wherever water is microscopically regulated for an unnecessary amount of energy to be absorbed by a plant, it is called the "wilting coefficient" or "permanent weakening level." Many plants cannot be reached for irrigation, which restricts the amount of water accessible for plant use. While clay keeps the primary water of all soil textures. Therefore, loams and silt loams square quantity vast volumes of water available to plants by the most beneficial soil textures [15].

11.1.3.3 Organic Matter

The soil organic matter endorses biotic growth. It acts as the nutritional resource to support earthworms along with new species. It has a large potential for H_2O absorption, a strong potential for moisture retention (between 80% and 90%), and provides other important nutrients for plants. It is the essential element that originates in soils at levels of roughly 1%–5%. It is derived from dead plants and animals and is itself highly capable of carrying on or providing the critical components and water for plant production. Soils that are peak in organic matter often have a higher CEC and, for that reason, are typically some of the most productive for the growth of plants. The organic matter often has an incredibly strong capacity to retain water on "plants available". This may improve the expansion ability of soils with reduced water-holding capability such as clay. The percentage of decayed organic matter in soils is routinely used as a productive soil indicator. However, excessive decomposition of natural materials may make it inaccessible for plant usage in time, producing recalcitrant carbon stores in soils [16, 17].

11.1.3.4 Gases

The next required component of the soil is gasses or dirt. As the air covers close places to water, it can make up between 2% and 50% of the amount of the earth. Gas is necessary for the respiration of the root or germ, which aids in the growth of plants. Carbon dioxide and nitrogen are also needed for the functions of underground plants, such as ammonia-fixing bacteria. When soils stay damp, root gas exchange may be prevented, contributing to plant death, which is a general issue behind floods [18].

11.1.3.5 Microorganisms

Microorganisms are the ultimate fundamental part of the soil; therefore, they grow in vast quantities.

Within the soil, but build up far less than 1% of the amount of the soil. Approximately, calculated is that one thimble topsoil may contain over 20,000 microorganisms. Most of such organisms are earthworms and nematodes, and the least of these are microbes, actinomycetes, algae, and fungi. Microbes are the primary untreated organic matter decomposers. Decomposers bring organic matter, soil, and air away to reprocess raw organic matter into humus, which is involuntarily abundant in plant nutrients. Many focused microorganisms, such as nitrogen-fixing bacteria, have a symbiosis with plants that permit plants to take away this necessary nutrient. These "nitrogen-fixing" plants are a major soil nitrogen aid and are important for soil growth over a moment of time. Mycorrhizas are complexes of fungi which outline strong-minded relations with plant roots. The fungus spreads toward the root of a plant, where the plant provides sugar to the fungus and in exchange. The fungus supplies roots to the plant in the company of water and permission for nutrients in the soil. Via the soil matrix, the complex network of hyphae spreads. The soil is dead, lacking microbes, and can be confined to sustain plant growth [19].

11.2 Concepts of Soil Chemistry

Soil chemistry is the latest limb of soil knowledge. The material is at the same level as important as that of soil physics, soil genesis and taxonomy, soil microbiology, soil fertility, soil mineralogy, and another branch of soil science. It is essential for each subdivision to carry a considerable amount of chemistry to explain their respective topics and issues. Many names are given for soil chemistry. Everyone recognizes concerning soil as a resource which fulfills fundamental prerequisite of humankind by sustaining a range of plants and other flora. It is complicated to relinquish a matchless explanation for it, as its observation is racist in the environment. Comprehensively, the soil will be visualized as a fine layer of the earth's surface, which acts as a natural platform for plant production. It has specific significances for various individuals, however. For a farmer, it's the part of the earth's surface where he will plow and produce food to meet family and animal requirements. To a civil engineer, the soil is the base of all development operations such as highways, houses, river systems, and drains. The origins of the petroleum cracking agents are sand and cement products for oil technologists. The soil is, therefore, a natural resource that needs proper maintenance to improve its quality. That helps to keep the environment clean and sustainable. In the last two centuries of scientific effort, the two models of soil have developed primary pedology. Does that treat soil as the ordinary creature, a chemically stained and synthesized result of existence and its categorization? The original definition is edaphology, which considers soil as natural surroundings for plants and study in terms of growth, regeneration, and creation [20, 21].

11.2.1 Soil Aeration

Soil biota refers to the shape which the soil took. It may occur as individual particles or as a mixture of primary particles from the soil-forming secondary particles called aggregates [22, 23]. What may be aggregate structure; it adds pores into the soil. Such pores are lined with water or dust and play a part in the development of plants. Plants breathe from their roots, and when there are no openings, and then the plants do not have air to breathe, so they will fail. Soil aeration, therefore, is one of the most important production factors in determining. Pore existence affects their bulk density (Db) and particle density (Dp). The ratio of soil particles and pore distance in soil may be calculated from either the bulk volume and particle density if the same units of calculation are represented in collaboration.

Amount of solid particles in soil = (bulk density/particle density) × 100

This percentage subtracted from the overall (100%) would give the percentage of pore difference. Hence, it will be the answer:

Percentage of pore space in soil = 100 – (bulk density/particle density) × 100

Soil pores are full of oxygen or carbon dioxide gases. These take part in the respiration of plants. Many essential results of soil aeration are oxidation-reduction processes that occur in soil, and this can assess the oxidation status of specific ions in the soil. If soil is very well oxygenated, there are oxidized forms that dominate Fe^{3+}, Mn^{4+}, NO_3^-, SO_4^{2-}, or poorly aerated reduced forms Fe^{2+}, Mn^{2+}, NH_4^+, and S^{2-} are found. These oxidation states influence the colors of the soil; red, yellow, and reddish-brown are present in oxidizing environments. Gray and blue play a large role because there is not enough oxygen. Crops are produced in the oxygenated form of acid soil appetite, although smaller levels of Fe^{2+} and Mn^{2+} are poisonous to the plants. The same is valid in drier areas, while Fe^{2+} and Mn^{2+} are favored [24–27].

11.2.2 Soil Colloid

A molecule with a diameter of 0.1–0.001 micrometer either is mineral or chemical. Colloids go into equilibrium in a solution due to excessive their tiny size—they move around for lengthy stretches of time, eventually coming out. Typical sources of soil colloids are the clay particles and synthetic plant products. They have a large relative surface area because of their limited size and have a charge, and they can adsorb cations. This is essential to cation exchangeability, but also maintaining the soil structure (binding particles) and its moisture-holding capacity (greater colloid content implies stronger water-holding capability).

11.2.3 Soil Water

The water inside the soil is known as the soil solution since it includes soluble substances (cations and ions) besides clay and organic product colloids suspended. Plants demand their nutrition from the soil water that, in one case, doesn't provide sufficient resources to sustain the plant's existence. Water has a significant function called surface tension, which affects its soil actions and can be determined by the volume of water in a capillary that falls. Because of the surface tension, water moves up into the Capillary tube while the downward force of gravity holds it back. In all the wet soils, capillary forces are found. However, the rate of movement and a rise in the height is less dependent on the pore size of the soil. The decline in the increase may be attributed to the several purposes: capillaries have a variable range, soil capillaries are not constant and smooth, capillaries have air obstruction, etc. Via the infiltration technique, water enters the profile and therefore passes via the profile via filtration. The ratios of micropores, mesopores, and macropores in the horizons of the soil often determine how rapidly water reaches and passes, although soil. Micro pores are the least minute opening in the soil that holds much of the moisture accessible to plants in the soil. In such pores, water becomes an order by the joint entry powers of sticking together and pulling together. Sticking together is the tendency of molecules of water to bind to a piece of dirt. Pulling together is the tendency of a single H_2O molecule (or other water molecules) to bind to itself. Meso pores are medium-sized pores created by various methods, such as Macro pores, rendering them the largest type of soil pore. These macropores are produced by animal burrowing (non-matrix pores). Others, by comparison, survive since the roots of the tree that once dominated the land have rotted away. Soil water has three energy forms [28–31]: kinetic, matric, and osmotic. Potential because of gravity relates to the force correlated with the Earth's water level; Earth's exterior water has a strong potential attributed because of gravity, yet gravitational potential after it hits the water table is nil. Matric capacity is the power inside the soil matrix (or profile) correlated with adhesion and bonding and is a detrimental potential as it is compared to gravity along with potential because of osmosis. Osmotic capacity tests substances dissolve inside soil; Greater the soluble chemical level, the fewer space water needs on the way to travel, but usually, this adds nothing toward the total flow of surface water, particularly in sandy soils. Potential for surface water is provided by the following simple formula:

$$\Psi T = \Psi G + \Psi M + \Psi O$$

Here,
ΨT = overall water potential of soil,
ΨG = potential attributed due to gravity,
ΨM = matric potential,
ΨO = osmotic potential.

This standing energy reproduces the number of energies necessary for extracting groundwater.

A badly organized soil may have a total porosity of as diminutive as 35%, although a better-designed soil with a similar type will have an overall porosity with 65%. In soil water movement and plant uptake, a well-structured soil will become more effective. Medium pores get more matric capacity (for the strength of adhesion and cohesion) than big pores. So, in fact, fantastic-textured soils possessing more nanopores keep sufficient water rather than coarse-consistency soils possessing a higher proportion of macropores. Upon entering the duration field capacity (FC) (ground soil may not be moist yet humid) is reached, much gravitational water becomes removed, bigger spaces are full of air, or liquid is an order in tiny pores. The persistent wilting point (PWP) is the level with which ground moisture for plant absorption is inaccessible (~−1500 kPa). Plants on PWP cannot tolerate H_2O heat, moreover may stay limp. Though, the smallest pores produce still slight quantities of absorbent material. Consideration is provided to water quality (AWC) as the soil water retained between FC and PWP.

11.2.4 Ion Exchange Property

Ion exchange is a reversible process. The soil can exchange its ion because of its colloidal properties. The water existing in the intermediate vicinity of clay and humus particles, which are both colloidal, is considered being water associated with exchangeable cations. The ion exchange reaction may cation exchange or anion exchange reactions. One reversible mechanism is the movement of ions. Because of its colloidal properties, the soil can exchange its ion. The liquid in the intermediate region of clay and humus fragments, all colloidal in composition, is known to be vapor connected with exchangeable cations. The reaction to the ion exchange can exchange reactions of cations or anions.

11.2.4.1 Cation Exchange Capacity (CEC)

The exchange power of positively charged ions is called, e.g., CEC. Ca^{2+}, Mg^{2+}, Na^{+}, K^{+}, NH_4^+, etc. The ion exchanges potential of soil is calculated by the proportional concentrations of different soil colloids and measured in centimoles of electric charge per unit of weight of a dry soil kitchen device. The monovalent cations are simpler to substitute than divalent cations or trivalent ones. The sequence of choice of certain tri, di and monovalent ions in cation exchange reactions are as defy:

$$Al^{3+} > Ca^{2+} > Hg^{2+} > K^{+} > Na^{+} > Li^{+}$$

The sandy soil has poorer cation exchange potential from clay soil, since the coarse-textured soil is usually lower in clay and humus content. The CEC of saturated soil decreases with pH; the load for clay mineral is fairly stable under pH 6.0, while the load on the mineral colloid rises beyond pH 6.0 due to the huge ionization of hydrogen from the activated hydroxyl group at the margins of the crystal. Although the CEC is normally weak at higher pH levels. The effects of cation exchange are important because it influences the physicochemical effects of soil. Hydrogen, potassium, and calcium are the main ions. Soil pH is influenced, for example, by the interaction of hydrogen ions. When transferable hydrogen predominates, then the surface becomes acidic. As a result, Al^{3+} ions are interchangeable and are possibly harmful to plants. The water is alkaline, with both the larger interaction of sodium ions. Calcium can develop safe, clearer particle aggregates [32, 33].

11.2.4.1.1 Different Variables That Affect CEC

1. The higher amount of clay in the soil indicates greater CEC. Other forms of clay (smectites, montmorillonite) have CEC better to others (e.g., kaolinite).
2. Higher concentrations of organic material in the soils create greater CEC.
3. Clay materials and organic material have a CEC that changes over pH. Because pH rises, so does the amount of negative loads onto the clay or organic material particles, so does the CEC.
4. Base saturation relates to the proportion of CEC locations shared by bases (usually Ca^{2+}, Mg^{2+}, K^+, and Na^+) in lieu of acidic soil ions (H^+ or Al3$^+$). Base saturation is also represented in percentage terms. Typically, the word transferable bases apply to Ca^{2+}, Mg^{2+}, K^+, and Na^+ adsorbed to CEC sites. Soils with large base saturations are called extra productive since plant nutrients are several of the "bases" that provide them. In addition, when the pH is above 6.5, the base saturation is 100%.

11.2.4.2 Anion Exchange Capacity (AEC)

Whereas positively valued cations adsorb negatively charged pages, for negatively charged anions, the same is valid. They adsorb into positively charged sites [33, 34]. It is the potential for an interchange of anions, AEC. Nutrients usually provided by anions are nitrogen (like NO_{3-}), phosphorus $\left(\text{like } HPO_4^{2-}\right)$, sulfur $\left(\text{like } SO_4^{2-}\right)$ chlorine (like Cl^-), boron $\left(\text{like } B_4O_7^{2-}\right)$, and molybdenum (like MoO^{4-}). Measurement: As with CEC (earlier in the thread), the Atomic power Commission is expressed as milliequivalents (meq) per 100 g of soil or centimoles (cmol) per kilogram. Much clay particles have only negative exchange sites, and they have CEC under favorable and high pH conditions, and often low pH AEC [35, 36]. Soil organic matter has both negative and positive exchange site; it also has CEC and in extremely poor pH situations would have Atomic power Department, a vast amount of best soils in the U.S. have pH far above the pH needed for AEC, so this route plays a marginal position in nutrient supply here.

11.2.5 Soil pH–Acidity and Alkalinity

Because of its huge impact on limitless chemical reactions involving critical plant nutrients, phytotoxic substances, and toxins, soil pH is known to be the "core component" of soil chemistry. pH is soil acidity or alkalinity analysis. It's often provided within 0 and 14 as a total. Soil pH values are given in Table 11.1. The pH is among the most significant characteristics concerned in plant production and also recognizing how quick reactions proceed in the soil. Crops generally prefer values of 5.5–8, but the value depends on the crop. Although pH does not influence plants directly, it influences plants' supply of various nutrients. The pH of the soil changes whether a mineral is dissolved in the soil solution or certain fewer soluble substances are produced, or if dissolved is prone to leach (e.g., nitrate). pH and soil microorganisms: in higher pH environments, soil microbes have less development. This may trigger them to take far longer to produce the required nutrients from organic matter, such as N, P, and S [37].

Soil pH differs by location, although acidic soils generally exist in wet environments, while alkaline soil is normally located within a region of minimal rainwater. Soils may turn acid by rocks weathering high in silica, creating organisms' acids, and removing acids on or after degrading natural matter. Alkaline soil is really the product of rock weathering like those of calcareous stone comprising significant quantities of calcium carbonate (salt) and salt dust inputs arising from drainage basin evaporation, especially in dry areas. Soil pH influences the solubility of nutrients and the speed of decomposition in soil, which has a significant impact on plant nutrient supply. A mildly acidic pH of approximately 6 and 7 provides plants with maximum nutrient supply, but there are exceptions. Many of the macronutrients fall within this spectrum. Understanding that soil pH status is essential for decision-making on crop management, as some growing plants need a fixed pH spectrum for optimum production. Soil pH can be evaluated by a pH meter in the laboratory or by quick test pH strips in the field [38, 39].

Table 11.1 pH and results.

Result	Extremely acidic	Highly acidic	Moderately acidic	Marginal acidic	Neutral	Mildly alkaline	Moderately alkaline	Highly alkaline
pH	3.5–4.4	5.1–5.5	5.6–6.0	6.1–6.5	6.6–7.3	7.4–7.8	7.9–8.4	8.5–9.0

11.2.5.1 Acid Soil Formation

The acidity of the soil is largely caused by hydrogen and aluminum ions, based on the level of acidity well into the soil. Aluminum is soluble at a quite low pH (<5) and hydrolyzes to $Al(OH)^{2+}$. Losing H^+ ions is accountable for acidity. At the pH range between 5 and 6.5, aluminum is transformed to $Al(OH)_2^+$ ions in the presence of OH^- which, after further hydrolysis from aluminum hydroxide and H^+ ions, and creates soil acidity further. Below, the sequential reactions are shown:

$$Al^{3+} + H_2O \rightarrow Al\,(OH)^{2+} + H^+$$

$$Al(OH)^{2+} + H_2O \rightarrow Al(OH)_2^+ + H^+$$

$$Al(OH)_2^+ + H_2O \rightarrow Al(OH)_3 + H^+$$

There are many other ways of formation of acid soils:

1. Flowing water and rainwater leads to the formation of acid soil.

$$Na^+ + \text{Soil-Water} \rightarrow H^+\text{-Soil} + NaOH$$

   Where alkali produced by running water become absorbed, or seeps downward, keeping acid soil behind.
2. Rainwater dissolves atmospheric carbon dioxide to produce carbonic acid and transforms soil into acid soil

$$Na^+\text{-Soil} + H_2CO_3 \rightarrow H^+\text{-Soil} + NaHCO_3$$

   In a polluted atmosphere, SO_2, SO_3, NO_2 gasses are found in the environment, dissolving in rainwater to produce H_2SO_3, H_2SO_4, and HNO_3, contributing to acid soils. As such acids are solid acids, in the presence of these; more acidic soils are produced than in the presence of carbonic acid.
3. Humus is the surface that hydrolyzes to create the salt, where it mixes ions with a surface to generate the salt soils.

$$Ca^{2+}\text{-Humus} + H_2O \rightarrow H^+\text{-Humus} + Ca(OH)_2$$

$$H^+\text{-Humus} + Na^+ \text{ - Soil} \rightarrow Na^+\text{-Humus} + H^+\text{-Soil}$$

4. In fertilizers, nitrogen is only absorbed by the plants in its nitrate type. Microbes transform the nitrogen in the fertilizers to nitrates as follows:

$$NH_4^+ + 2O_2 \rightarrow 2H^+ + NO_3^- + H_2O$$

These hydrogen ions experience an interchange of ions to create the soil acidic. Inorganic fertilizers, then, cause the soil acidic.
5. Many oxidation reactions occurring in the soil are correlated with H+ ion formation as follows:

$$Mn^{2+} + 2H_2O \rightarrow MnO_2 + 4H^+ + 2e^-$$

$$2Fe^{2+} + 3H_2O \rightarrow Fe_2O_3 + 6H^+ + 2e^-$$

$$2Fe_2O_3 + 3H_2O \rightarrow 3Fe_2O_3 + 2H^+ + 2e^-$$

Through the elimination of these H+ ions, the equilibrium changes to the right side.

11.2.5.2 Acid Soil Benefits

Acid soil produced by either of the above-mentioned processes acts as strong acids in such a way that pH is poor only in partial conversion to acid soil. Acid soils provide other properties for planting:

1. All metal oxides and hydroxides perform easy dissolution, making a metal ion readily available for plant growth as nutrients.
2. Carbonates break down in the soil to generate further CO_2, and that is important to plant production.
3. Acid soil makes phosphate fertilizers quickly available for plants to use.
4. Acid soil breaks down the mica and illite found in rock. So that plants would have more potassium ion availability to expand.

Soil acidity could have adverse effects on plants because of influences on enzymatic shifts, decreased nutrient supply such as copper, zinc, retarded microbial behavior, certain nutrient leaching, etc.

11.2.5.3 *Liming-Decreasing Soil Acidity*

Soil acids may be handled in two forms, either by planting crops appropriate for a specified pH or by manipulating soil with chemicals that will overcome soil acidity. Former is dangerous, costly, and persistent field production of arable crops in the wetland area would worsen soil acidity. So the usage of acid-producing fertilizers on acid soil would make the soil unproductive as further acidosis develops [40]. Liming soil is necessary to mitigate the adverse impacts of soil acidity and aid in the management of acid soils. Acid soils are made more appropriate for farming by liming, which increases soil pH. Soil acidity is usually reduced by adding Ca Mg carbonates, oxides, or hydroxides, called to a farm tones.

Liming

- It reduces the solubility and supply of certain plant nutrients and decreases the amount of Ca and Mg exchangeable base status.
- Influences plant nutrient intake by generating unique antagonistic results.
- Reduces the aluminum and manganese toxic content by neutralizing the effect.
- Enhancing soil composition and promoting root spread.

In particular, the direct and indirect effects of lime application on acid soils are enhanced crop production, although they vary in magnitude under specific soil conditions and cropping systems. In acid soils, legumes are more sensitive to the application of lime. The key explanation for the differential answer to lime is because the soil reaction comparisons of various crops are not equivalent to normal growth.

11.2.5.4 *Alkaline Soil Formation*

Soil alkalinity perhaps because of increased base saturation, especially with calcium and sodium and free carbonates. Soil alkalinity may be two-way:

1. In the base, soluble carbonates hydrolyze according to the corresponding reaction to make the base alkaline

$$Na_2CO_3 + H_2O \rightarrow 2Na^+ + 2OH^- + H_2CO_3$$

2. Soil that produces excess Ca or Mg carbonates undergoes hydrolysis to alkaline the soil as H_2CO_3 is a poor acid.

$$CaCO_3 + 2H_2O \rightarrow Ca^{2+} + 2OH^- + H_2CO_3$$

11.2.5.4.1 Disadvantages of Alkaline Soil

Many plants will integrate nitrogen in the nitrates, so oxidation of ammonium fertilizers will occur only in an acidic atmosphere; plants in an alkaline atmosphere do not have nitrogen accessible. Both metal ions are precipitated as hydroxides, which are not accessible to the plants as nutrients. Phosphates have are precipitated and do not arrive with nutrients. To put it another way, plant growth is sluggish in an alkaline atmosphere.

Acknowledgments

Dinesh Kumar is grateful for financial assistance from DST, New Delhi, for this (sanctioned video project Approval Order F. No. DST/TM/WTI/WIC/2K17/124(C).

References

1. Price, G., *Australian Soil Fertility Manual*, Fertilizer Industry Federation of Australia, Inc. & CSIRO, Collingwod, Australia, 2006.
2. Abruna, F., Vicente Chandler, J., Pearson, R.W., Effects of Liming on Yields and Composition of Heavily Fertilized Grasses and on Soil Properties Under Humid Tropical Conditions. *Soil Sci. Soc. Am. J.*, 28, 657–661, 1964.
3. Curtin, D. and Syers, J.K., Lime-induced changes in indices of soil phosphate availability. *Soil Sci. Soc. Am. J.*, 65, 147–152, 2001.
4. Lindsay, W.L., *Chemical Equilibria in Soils*, John Wiley and Sons Ltd, Hoboken, NJ, USA, 1979.
5. Sparks, D.L., *Environmental Soil Chemistry*, Elsevier, Amsterdam, The Netherlands, 2003.
6. Sims, J.T., Cunningham, S.D., Sumner, M.E., Assessing soil quality for environmental purposes: roles and challenges for soil scientists. *J. Environ. Qual.*, 26, 20–25, 1997.
7. Sojka, R.E. and Upchurch, D.R., Reservations regarding the soil quality concept. *Soil Sci. Soc. Am. J.*, 63, 1039–1054, 1999.
8. Sojka, R.E., Upchurch, D.R., Borlaug, N.E., Quality soil management or soil quality management: performance versus semantics, in: *Advances in Agronomy*, vol. 79, Sparks, D.L. (Ed.), pp. 1–68, 2003.

9. Schlatter, D., Kinkel, L.L., Thomashow, L.S., Weller, D.M., Paulitz, T.C., Disease suppressive soils: new insights from the soil microbiome. *Phytopathology*, 107, 1284–1297, 2017.
10. Dexter, A.R., Advances in characterization of soil structure. *Soil Tillage Res.*, 11, 199–238, 1988.
11. Rutgers, M., Schouten, A.J., Bloem, J., van Eekeren, N., de Goede, R.G.M., Akkerhuis, G.A.J.M.J.O., van der Wal, A., Mulder, C., Brussaard, L., Breure, A.M., Biological measurements in a nationwide soil monitoring network. *Eur. J. Soil Sci.*, 60, 820–832, 2009.
12. Bouma, J., Soil science contributions towards Sustainable Development Goals and their implementation: linking soil functions with ecosystem services. *J. Plant Nutr. Soil Sci.*, 177, 111–120, 2014.
13. Letey, J., Sojka, R.E., Upchurch, D.R., Cassel, D.K., Olson, K.R., Payne, W.A., Petrie, S.E., Price, G.H., Reginato, R.J., Scott, H.D., Smethurst, P.J., Triplett, G.B., Deficiencies in the soil quality concept and its application. *J. Soil Water Conserv.*, 58, 180–187, 2003.
14. Cooper, E.L., *Agriscience: Fundamentals & Applications*, 2nd. Ed., Delmar Publishers, Albany, New York, 1997.
15. Essington, M.E., *Soil and Water Chemistry: An Integrative Approach*, 2nd ed., Routledge, Abingdon, UK, 2015.
16. Doran, J.W., Sarrantonio, M., Liebig, M.A., Soil health and sustainability. *Adv. Agron.*, 56, 1–54, 1996.
17. Patzel, N., Sticher, H., Karlen, D.L., Soil fertility - phenomenon and concept. *J. Plant Nutr. Soil Sci.*, 163, 129–142, 2000.
18. Schlesinger, W.H. and Andrews, J.A., Soil respiration and the global carbon cycle. *Biogeochemistry*, 48, 7–20, 2000.
19. Emerson, W.W., Emerson dispersion test, in: *Soil physical measure- ments and interpretation for land evaluation*, N. McKenzie, K. Coughlan, H. Cresswell (Eds.), pp. 190–199, CSIRO Publishing, Collingwood, Vic., 2002.
20. Arshad, M.A., Lowery, B., Grossman, B., Physical tests for monitoring soil quality, in: *Methods for assessing soil quality*, Soil Science Society of America Special Publication 49, J.W. Doran and A.J. Jones (Eds.), pp. 123–142, SSSA, Madison, WI, 1996.
21. Blanco-Canqui, H. and Lal, R., Soil and water conservation, in: *Principles of Soil Conservation and Management*, pp. 1–19, Springer, New York, 2008.
22. Emerson, W.W., A classification of soil aggregates based on their coherence in water. *Aust. J. Soil Res.*, 5, 47–57, 1967.
23. Tisdall, J.M. and Oades, J.M., Organic matter and water stable aggregates in soils. *J. Soil Sci.*, 33, 141–163, 1982.
24. Carreira, J.A., Niell, F.X., Lajtha, K., Soil nitrogen availability and nitrification in Mediterranean shrublands of varying fire history and successional stage. *Biochemistry*, 26, 189–209, 1994.
25. Debano, L.F., Neary, D.G., Ffolliot, P.F., *Fire's effects on ecosystems*, John Wiley and Sons, Inc., New York, 1998.

26. Edwards, P., Schoonover, J.E., Williard, K.W.J., Guiding principles for management of forested, agricultural, and urban watersheds. *J. Contemp. Water Res.*, 154, 60–84, 2015.
27. Fisher, R.F. and Binkley, D., *Ecology and Management of Forest Soils*, 3rd edition, John Wiley and Sons, Inc., New York, 2000.
28. Karlen, D.L., Mausbach, M.J., Doran, J.W., Cline, R.G., Harris, R.F., Schuman, G.E., Soil quality: A concept, definition, and framework for evaluation (A Guest Editorial). *Soil Sci. Soc. Am. J.*, 61, 4–10, 1997.
29. Kimmins, J.P., *Forest Ecology. A Foundation for Sustainable Management*, 2nd edition, Prentice Hall, Inc., Upper Saddle River, New Jersey, 1997.
30. Abdollahi, L., Hansen, E.M., Rickson, R.J., Munkholm, L.J., Overall assessment of soil quality on humid sandy loams: effects of location, rotation and tillage. *Soil Tillage Res.*, 145, 29–36, 2015.
31. Kay, B.D., Influence of aggregate diameter, surface area and ante- cedent water content on the dispersibility of clay. *Can. J. Soil Sci.*, 70, 655–671, 1990.
32. Rengasamy, P. and Churchman, G.J., Cation exchange capacity, exchangeable cations and sodicity, in: *Soil analysis: an interpretation manual*, K.I. Peveril, L.A. Sparrow, D.J. Reuter (Eds.), pp. 147–157, CSIRO Publishing, Collingwood, Vic, 2001.
33. Osaki, M., Watanabe, T., Tadano, T., Beneficial effect of aluminum on growth of plants adapted to low pH soils. *Soil Sci. Plant Nutr.*, 43, 551–563, 1997.
34. Zelazny, L.W., He, L., Vanwormhoudt, A.N., Charge analysis of soils and anion exchange, in: *Methods of Soil Analysis Part 3—Chemical Methods*, D.L. Sparks, and *et al.* (Eds.), pp. 1231–1253, SSSA, ASA, Madison, WI, USA, 1996.
35. Hingston, F.J., A review of anion adsorption, in: *Adsorption Inorganic Solid Liquids Interfaces*, pp. 51–90, Ann Arbor Science, Ann Arbor, MI, USA, 1981.
36. Scharenbroch, B.C., Lloyd, J.E., Johnson-Maynard, J.L., Distinguishing urban soils with physical, chemical, and biological properties. *Pedobiologia*, 49, 283–296, 2005.
37. Agboola, A.A. and Corey, R.B., The relationship between soil pH, organic matter, available phosphorus, exchangeable potassium, calcium magnesium and nine elements in the maize tissue. *Soil Sci.*, 115, 367–375, 1973.
38. Abdulgawad, G., Bashour, I., Elmadhon, Y., Sink holes formation in some salt-affected soils and their origin. *Libyan J. Agric.*, 6, 159–164, 1977.
39. Abrukova, L.P. and Isayev, V.A., Structural-mechanical properties of gray-brown gypsiferous soils. *Sov. Soil Sci.*, 15, 6, 90–100, 1983.
40. Barrow, N.J., Further investigations on the use of lime on established pastures. *Aust. J. Exp. Agric.*, 5, 442–449, 1965.

12

Essential Soil Functions for Enhanced Agricultural Productivity and Food Production

Charles Oluwaseun Adetunji[1]* and Osayomwanbo Osarenotor[2]

[1]Applied Microbiology, Biotechnology and Nanotechnology Laboratory, Department of Microbiology Edo University Iyamho, Edo State, Nigeria
[2]Department of Environmental Management and Toxicology, Faculty of Life Sciences, University of Benin, Benin City, Edo State, Nigeria

Abstract
Soil biological indicators have been identified as essential factors that could enhance the normal function of soil. They also determined the sustainability of essential ecological factors that supports sustainable agriculture and increase in food production. They include number and type of macro and microbiota and microbial biomass, while soil biochemical indicators are soil enzymatic activities, organic matter decomposition, and cycling of bio-nutrients which are needed to support plant growth. Therefore, this chapter discusses some recent information significant ecological factors that could boost agricultural production, maintenance of a cleaner environment and increase in food production that could help to med the demand of the ever increasing production. Detailed information on the relationship between CO_2 levels in the atmosphere and global carbon cycle and soil carbon as a typical example of soil function, relationship between soil organic content and soil carbon sequestration, essential soil biological processes and their relationship with soil pH, and microbial ecophysiological indicators and activities of soil enzymes were also discussed.

Keywords: Agriculture, food production, soil biological processes, soil respiration, soil function, climate changes, soil enzymatic activities, soil organic content, soil carbon sequestration

**Corresponding author*: charliguitar@yahoo.com; adetunjicharles@gmail.com; adetunji.charles@edouniversity.edu.ng

Inamuddin, Mohd Imran Ahamed, Rajender Boddula and Tariq Altalhi (eds.) Applied Soil Chemistry, (215–234) © 2021 Scrivener Publishing LLC

12.1 Introduction

The ability of soil to maintain the quality of the environment and promote plant and animal health is known as soil function [47]. Soil functioning depends on the ability of the soil to sustain biological activities, and hence, soil biological and biochemical properties are the best indicators of soil quality. This is because they are most sensitive to natural and biological disturbances to soil.

Soil biological indicators include number and type of macro and microbiota and microbial biomass, while soil biochemical indicators are soil enzymatic activities, organic matter decomposition, and nutrient cycling [15]. Various factors affect soil function, thereby resulting in poor soil quality and reduction or loss of soil fertility. Ananbeh *et al.* (2019) [1] used soil enzymatic activities to access the recovery of an abandoned soil. Jung *et al.* (2016) [22] used nitrogen cycling as an indicator of soil quality deterioration in diesel contaminated soil. Bonanomi *et al.* (2016) [7] studied the organic matter decomposition rate as an indicator of effect of organic farming on agricultural soil.

Over tillage of soil has been reported to affect soil function negatively by several authors [1, 60] who also reported that change in climatic conditions overtime can affect functions such as soil organic carbon transformation and nitrogen cycling. Pollution of soil with petroleum hydrocarbon has also been reported to cause loss of soil functioning, Jung *et al.* (2016) [22] showed that the diversity of bacteria was greatly reduced in diesel contaminated soil with resultant reduction in nitrogen cycling. The use of pesticide in agriculture is largely inevitable; however, it also has its detrimental on soil function. Several authors have reported that application of pesticides in farms can affect vital soil function [59].

Therefore, this chapter intends to explain the detailed information on the necessary soil functions that could led to increase in agricultural production and increase in food production. Essential soil biological processes and their relationship with soil pH and microbial ecophysiological indicators and activities of soil enzymes were also discussed.

12.2 Effect of Pesticides on Soil Function Indicator

Pesticides are of various types depending on the target; they include insecticides, rodenticides, fungicides, and nematicides [59]. They are used in agriculture to protect the plant from their action of their targets thereby increasing crop yield. However, they have been shown to affect non-target beneficial biological entities in the soil, thereby resulting in poor soil quality or loss of soil function [4]. A good pesticide ideally should dissipate after performing its action; however, this may not be the case as they remain in the soil. This factor coupled with the soil serving as sink for pollutants results in its persistence in the soil [33].

Wang *et al.* (2018) [57] studied the effect of Azoxystrobin, a novel strobilurin fungicide on the biological and biochemical activities of soil. Cabisol soil samples were spiked with azoxystrobin solution in acetone at 0.1 and 1.0 mM and 10 mg/kg of dry soil, thus twice the recommended concentration of the fungicide, followed by incubation in the dark for 28 days. After the incubation period, soil samples from the various treatment groups were analyzed for soil function using oil microbial population, soil nasal respiration, as well as soil dehydrogenase and urease activity. Results of soil microbial population showed that azoxystrobin application showed significant effect on bacteria and actinomycetes population, this was evident in the observed strong negative correlation between the microbial population and the concentration of the fungicide. The authors therefore concluded that the fungicide had effect on the soil microbial population and which could have impact on the soil quality. This is so considering the fact that soil microorganisms play crucial role in soil organic matter decomposition; they therefore suggested that further studies should investigate the corresponding effect of the fungicide on soil organic matter decomposition and transformation. They also pointed out that no significant effect was observed in the fungi population and attributed this to the low pH of the soil (5.22) which can support fungi growth; they therefore concluded that response of soil microbes to fungicide could be selective depending soil characteristics. Wang *et al.* (2018) [57] also showed that azoxystrobin-treated soils had significant reduction in soil basal respiration but concluded that it was not influenced by soil type unlike soil microbial population. Dehydrogenase activity also reduced drastically and was also negatively correlated with concentrations of azoxystrobin used. Their result also showed that dehydrogenase activity in the soil was influenced by soil types, as indicated that fungicide effect on dehydrogenase was also influenced by soil type.

Yao *et al.* (2006) [63] tested the effect of insecticide, acetamiprid, on soil functions in a microcosm experiment. They used soil quality indicators: urease, phosphatase, and proteinase activities as well as soil respiration. Varying concentration of the insecticide, 0.5, 5.0, 50 mg kg^{-1} were added to 1,500 g of dried soil in a plastic pot. The experimental soils were incubated for 50 days. Their results showed that there was significant effect of acetamiprid on dehydrogenase activity after 3 weeks of incubation. The activity of the enzyme was lower in the treated samples when compared with the control. Although there was significant effect of acetamiprid application on soil urease activities, their result however showed that this effect was not govern by concentration as no significant effect was observed at varying concentration of the insecticide. Significant reduction was also observed in the activity of soil proteinase enzyme after prolong treatment with acetamiprid, with the most change seen after 4 weeks of incubation in the 50 mg kg^{-1} treated soil where activity reduced by 0.45-fold after 28 days of application.

Mukherjee *et al.* (2016) [34] carried out pot experiment to determine the effect of two herbicides persistence on soil functions, using microbial indicators. Concentrations of 5.7 and 57 mg/kg each herbicide was applied to 250 g of each soil type followed by incubation for 28 days. After the incubation period, soil function was accessed using microbial biomass carbon, substrate induced respiration, microbial metabolism quotient, β-glucosidase, and fluorescein diacetate hydrolysing activity. The results showed that significant reduction in microbial biomass carbon in both soil types at field recommended and 10-fold increase in field recommended concentrations for both herbicides up to 30 days. At both concentrations, persistence of the insecticides was correlated with decrease MBC. However, concentration of the insecticides has strong influence on MBC; strong inhibition of MBC was observed for the higher concentration. The authors also reported that inhibition of soil microbial biomass at field recommended concentration was temporary as it increased after 30 days once the decay of the insecticides begin. Fluorescein diacetate hydrolysis was utilized to validate the level effect of the pesticide on soil microorganism activities. It showed that AL soil treated with florasulam and 10-fold florasulam had significant lower activity than control soil up to 10 days. For SAL soil significant lower activity was recorded up to 15 days for both FR and 10-fold FR treated soils, respectively. They attributed the effect of the insecticide to soil type and concentration. Fluorescein diacetate hydrolysis assay is a higher enzymatic assay that necessitates the activities of these microbial enzymes at the same time [46].

Fierer *et al.* (2006) [15] studied effect of pesticide contamination on enzyme actions. Their study focused on the use of active ingredients of ten pesticides (2,4DAryloxyacid, Carbaryl, Mancozeb, Glyphosate, Azinphos-methyl, Parathion-methyl, Atrazine, Prometryne, Diuron, and Linuron) in microcosm study at varying incubation time of up 10 months. They also used the biology Ecoplates to study functional bacterial diversity. The authors discovered that high variability in the activities of phenol oxidase, arylamidase, and β-glucosidase. Phenol oxidase was the only enzyme that showed early increase in activity during the experiment. Arylamidase and β-glucosidase activity returned back to its background levels at the end of the experiment.

Arora *et al.* (2019) [5] studied the effect of pesticide application by farmers on paddy fields in India. The study employed the use of soil bacterial population, microbial biomass carbon, and dehydrogenase enzyme as indicators of soil function. Fresh soil samples were collected from rhizosphere of paddy in 20 locations which were selected based on the known history of pesticide application. Their findings showed that bacterial population dropped from 6.2×10^5 CFU/g to 1.8×10^4 CFU/g in locations with high pesticide applied. This was contrary to significantly increase in bacterial population observed in locations with low pesticide usage, where bacterial population was enhanced from 5.4×10^4 CFU/g to 4.1×10^5 CFU/g. Result from field study also showed that there was variation in MBC from 24 to 118 mg/kg in soils where excess pesticides were used while the range of MBC in soils with safe use of pesticides was 60 to 182 mg/kg. Soil microbial biomass is an indicator of pesticide degradation capacity [56]. Incubation of soil after 90 days with 1 ppm of four commonly used pesticides, chlorpyriphos, 2,4-D, carbendazim, and carbofuron, was also studied by these authors. The experiment was carried out in randomized block design with three replicates; the result showed that there was high reduction in dehydrogenase activity in soil by 42.97%, 58.24%, 28.68%, and 43.96% in soil treated with chlorpyriphos, 2,4-D, carbofuron, and carbendazim, respectively compared to control soil.

12.3 Soil Organic Matter

Soil carbon storage has been recognized as an essential component of ecosystem service that emanate as a result of relationship of several ecological processes. It has been discovered that numerous human activities which has influence on these could led to the loss of carbon or enhanced carbon storage. Organic matter has been recognized as a major component of soil that could affect its biological, physical, and chemical qualities backing momentously to its backing momentously which consequently affect human society. The merits of soil organic matter entails enhanced soil quality by improving the level of water retention, nutrient level which normally increases the level of agricultural productivity and maintenance of a cleaner environment, Moreover, SOM also enhances the level of soil structure and decrease the rate of soil erosion, which eventually led to enhance the level of water quality available in the surface water and ground water all this will improve the level of food security and decrease in negative influences to ecosystems [30].

It has been revealed that adequate knowledge of soil productivity could be linked to the soil organic matter content. This implies that whenever there is decrease in the level of SOM might lead to have a detrimental effect on the whole planet as well as ecosystem. Moreover, the soil disturbance could also result to decrease in leaching and enhanced erosion from soil which could result to high level of eutrophication and could eventually led to algal blooms within most especially in the coastal and aquatic ecosystem which eventually led to dead zones in the ocean. The restoration of the right level of organic matter in the soil necessitates proper understanding of the ecological process which is crucial for SOM storage. Adequate restoration methodology could help in the proper restoration of terrestrial ecosystem functions [38].

12.4 Soil Organic Carbon

The soil organic carbon entails essential soil microbes which entail fungi and bacteria, fecal material, decomposing material derived from animal and plant tissues, and the eventual product obtained from their decaying products. SOM consists of the mixture of heterogeneous materials that consist of decomposed materials from plant residues which eventually turns to humus. The soil organic carbon level could be associated with the level of organic matter that is present in soil which also determines the level of SOC.

Typical examples of SOC level results from the association of numerous ecosystems include respiration, decomposition, and photosynthesis which are essential major component. Furthermore, photosynthesis could be referred to as the fixation of atmospheric CO_2 into the biomass of plant. The rate of SOC could be linked or associated to the rate of root biomass of plant which also entails decomposition of plant shoots. The total soil carbon could be measured directly from the decomposition of the root of plant and the indirect movement of carbon-enhanced substances derived from the root to these beneficial soil microorganisms. A typical example of this is soil mycorrhizae that develop a symbiotic relationship with the root of the plant. The fungi gives the plant essential nutrients such as phosphorus while the root provides energy inform of carbon. Decomposing biomass that is available in the soil majorly from the biodegradation of beneficial soil microorganisms could led to carbon loss in the form CO_2. Also, it has been observed that leaching and erosion might lead to the decrease and eventual removal of carbon losses into the ground water. Moreover, when the level of carbon output and carbon input are balance with each other, there is no net alteration in soil organic content level. Also, when the input of carbon derived from photosynthesis exceeds C losses, SOC level increases within the period of time.

Furthermore, it has been observed that respiration rate, photosynthesis, and decomposition rate is affected comparatively by climatic influences most essential soil moisture and soil temperature. Atypical illustration of this could be found around the northern latitudes most especially the cold wet climates where the level of photosynthesis is greater than the level of decomposition leading to high level of soil organic carbon. The arid regions possess minimal level s of SOC which could be linked to high rate of minimal primary production while the topics normally possess intermediate SOC levels which might be linked to decomposition during warm temperature, primary production, and the presence of moisture level. Also, high level of productivity was experienced at the most temperate ecosystem which possesses a very high level of primary productivity majorly during the rest of the year which eventual delay the rate of decomposition and allows the organic matter to build up slowly. Also, it has been highlighted that climatic conditions also influence the level of soil carbon which could also affect the rates of soil organic carbon.

12.4.1 Relationship Between Global Carbon Cycle and Soil Carbon as a Typical Example of Soil Function

The amount of carbon available in the soil indicated the level of carbon available in the terrestrial ecosystem on the earth. It has been discovered the summation of carbon that exist in the terrestrial environment could be approximately valued at 3,170 gigatons. Also, 1 GT represent 1 pentagram which is equivalent to 1 billion metric tons, through which, only 80% which represent 2,500 GT is available in the soil [25]. The level of soil carbon could be inorganic (950 GT) or organic (1,550 GT) which represent the total number of carbonate materials and elemental carbon which entails gypsum, calcite, and dolomite [25]. It has been observed that the level of carbon available in animal and plants is very minimal when compared to that of the soil. The soil carbon pool has been identified to be nearly 3.1 times when compared to the carbon pool available in the atmospheric with the value of 800 GT [39] while the highest obtained at the ocean could be estimated at 38,400 GT [18].

12.4.2 Relationship Between CO_2 Levels in the Atmosphere and Soil Carbon Content

Moreover, the level of carbon generated into the atmosphere mainly through the process of deforestation entails the release of carbon generated from the process of decomposition of the aboveground plant biomass which eventually makes the carbon level in the soil to be removed from the putrefaction of the soil organic matter. Also, there is a nexus between the oceanic carbon storage rates at the marine ecosystems and global nutrient cycles [9].

12.4.3 Relationship Between Soil Organic Content and Soil Carbon Sequestration

Soil carbon sequestration could be referred to as the process which involves withdrawing of CO_2 derived mainly from the atmosphere and their eventual storage in the soil in form of carbon pool. Photosynthesis plays a major role in the buildup of the soil organic carbon [25, 26]. Therefore, there are several factors which could influence the level of soil organic carbon content such as topographic heterogeneity, climatic controls, current land management approaches, and historic land use patterns.

Enhanced increase in the level of global temperatures and atmospheric CO_2 may lead to different effect on soil carbon feedbacks by controlling the rate of carbon losses and the degree of the photosynthetic through decomposition and respiration. It has been documented that plants that were grown in enhanced CO_2 concentrations possess the tendency to fix more carbon mainly through the process of photosynthesis which might eventually led to the generation of improved biomass [12]. Also, the increase in the biomass might be linked to the enhanced rate in the level of plant respiration mainly through the plant root [19] or through enhanced microbial degradation of soil organic matter mainly through microbial activities [64].

Also, enhanced temperature might improve the plant productivity which might have influence on the level of carbon balance [28] while increase in the level of temperature might lead to enhancement of soil organic matter which might consequently generate more CO_2, leading to constructive responses on climate change [41]. Furthermore, landscape heterogeneity may affect some ecosystem processes and carbon sequestration capacity available in the soil, and some of these processes may vary such as soil erosion, soil temperature, rainfall infiltration, and deposition of sediment. All these have a considerable effect on the carbon loss rate and carbon input which might affect the level of soil organic carbon content in different topographic gradients [53]. A typical example of this could be found in a sloppier area which affects the nutrients and soil moisture level which consequently affect the level of soil carbon contents and all these might eventually affect the level of carbon sequestration capability across landscapes. Apart from carbon sequestration which may affect the level of soil functionality and soil organic carbon, there are some other management practices that could affect the level of soil organic carbon such as reduced tillage, use of cover crops, addition of organic manure or amendments, and control of erosion.

Soil pH has been identified as one of the crucial components of soil function that is responsible for the level of soil fertility. This also determines the supply of necessary nutrients and the function of the soil and its features most especially in an ecosystem. The biogeochemical processes have been recognized a broad discipline that embrace inter-multidisciplinary involving soil biogeochemistry, soil microbiologist, and soil scientist. Most of these scientist studies biogeochemical processes, which focuses on the several soil functions in an ecosystem which has a stronger relationship with soil biogeochemical processes which form a stronger tide between the geological, biological, and chemical [10]. Soil has been identified as a necessary element of life support systems because it ensure adequate delivery of numerous function in the ecosystem which includes soil fertility, carbon storage, food production, and water regulation [14, 21].

Millennium Ecosystem Assessment (2005) [21] stated that the adequate provision and regular functioning have a greater impact on the components of the well-being in the contest of their adequate health and safety, good social relations, and provision of good materials for life. Furthermore, the pH of the soil has been shown to have a greater effect on the soil biogeochemical processes. This shows that the soil pH could be refers to several types of soil characteristics which entails physical and chemical biological features which could influence the plant biomass yield and their growth [8, 32].

Moreover, soil pH also influences other soil functions such as the removal of some basic cations which includes Na, Ca, K, and Mg whenever they are liberated during the process of weathered minerals. The separation of the CO_2 in soil water which leads to the formation of H^+ ions; the humic residues obtained during the process of humification of soil organic matter generates phenolic groups and high-density carboxyl which eventually forms H^+ ions; nitrification of NH_4^+ to NO^{3-} produces H^+ ions; the uptake of nitrogen in animal products and plant; the inputs obtained from acidic rain and the absorption of nutrient by plant [60].

Therefore, it very obvious that pH regulates the biology of the soil and other soil functions as well as other biological processes. Moreover, there is stronger relationship between biogeochemical processes in terrestrial ecosystems and the soil pH. Furthermore, some recent advances in researchers have shown that soil pH play a crucial role in the regulation of numerous soil processes which involves the breaking down of xenobiotics substances when exposed to the effect of soil pH. The three phases of the soil include their fate after biodegradation, translocation, and transformation. This also shows that soil pH will determines the fates of substances in the soil environment which have effect on the level and rates of nutrients recycling and their eventual availability for production of crop, movement of hazardous materials in the surroundings maintenance of a cleaner environment, and bioremediation of heavily contaminated environment [36].

12.4.4 Essential Soil Biological Processes and Their Relationship With Soil pH

12.4.4.1 Microbial Ecophysiological Indicators

Ecophysiology has been recognized as an interlinkage that exists between the functioning of cell physiological when subjected to the effect of numerous environmental factors [2]. It has been stated that the utilization of the metabolic quotient (qCO_2) as an index to indicate the effectiveness of organic substrate usage mainly through the help of beneficial soil microorganism in specific soil conditions [2, 43].

It has been realized that reduction in the microbial community respiration could ensure the availability of numerous biomass production which could ensure better biomass per unit [2]. This shows that metabolic quotient could be refers to as the cell-physiological entity which shows the level of alteration when subjected to soil environmental conditions [2]. This might be regulated by the soil pH [6].

The soil pH has been identified as an essential driving force which perform essential function in the microbial ecophysical indices stems which also determines their influence on the microbial ecosystem and participates in the maintenance of the community which constitute a significant predictors of the metabolic quotient [36, 37]. It has also been discovered that metabolic quotient was discovered that a value of 2.5 greater in a lower pH soils in comparison to pH soils which is neutral in nature [6]. It has been stated that the optimum pH essential for effective microbial action varies from 5.5 to 8.8 [15, 43] while the soil respiration habitually rises with soil pH to an optimal level after such microbial action [44].

Moreover, there is a correlation with the microbial biomass N and C contents, which could be found to be greater above pH 7 [43]. Also, it has been stated that in a lower pH conditions, fungal respiration normally signifies a greater bacterial respiration [2]. This might be linked to the fact that because fungi could adjust to acidic soil conditions when compared to bacteria.

12.4.5 Soil Enzyme Activities and Their Relationship With Soil pH

Enzymes that are generated extracellular through the action of beneficial soil microorganisms for effective biogeochemical cycling of some essential nutrients that are necessary for effective development of plant [54]. It has been realized that soil pH constitutes a significant aspect which is necessary for the appropriate operational and regulation of the activity of the soil enzyme available in the soil and they may also affect the rate of enzyme production of these soil microorganisms [23]. Numerous beneficial enzymes are involved in the biological system which performs an effective function in the biotransformation of numerous substances.

These soil enzymes are developed from numerous sources with various level of stabilization majorly of solid surfaces. These enzymes possess that capability to biodegrade these substrates most especially when utilized at an optimum pH condition [54]. A typical example of this is the phosphorus enzymes which includes alkaline phosphatase enzymes which act best within the rages of pH of 8.5–11.5 and acidic phosphatase which act best within the pH of 3–5.5 [23]. Moreover, Stursova and Walker (2006) [52] discovered that organophosphorus hydrolase has the highest enzyme action at a greater pH [23, 50, 51].

It has been highlighted that alteration in the microbial community composition could affect potential the effect of enzyme generation if the numerous microorganism groups require minimum nutrients concentration which area needed for the construction of biomass or possess enzymes which varies in affinity for nutrients [4].

12.4.6 Biodegradation of Toxic Substances by Soil Microorganisms and Their Relationship With Soil pH

The process of biodegradation could be referred to as the chemical dissolution of inorganic and organic pollutant by biological agents or microorganisms [11, 42]. It has been discovered that soil pH has a greater influence on the rate of soil biodegradation most especially through it influence on microbial community and diversity, microbial activity, and the action of enzymes that play a crucial role in the biodegradation poisonous substances. For instance, slightly acidic or alkaline influences the rate of biodegradation while acidic environment could prevent the rate of biodegradation [42, 52]. It has been observed that the pH values between 6.5 and 8.0 could influence the rate of oil degradation [55].

12.4.6.1 Water Holding Capacity

It has been documented that there is a correlation between soil organic matter and the amount of soil holding capability. It has been observed that the level of water that area made available to plant varies from 10 and 1,500 kPa which increases for a sandy soil with cumulative soil organic carbon which later increases for clayed soil [45, 59]. The level of water holding capacity in the soil could be linked to the total amount of soil organic matter which also determined the variation that is possible in numerous pedotransfer functions.

12.4.6.2 Soil Erodibility

Soil erodibility has been recognized as one of the major factors that could determine the capacity of water erosion most especially for rainfall erosivity, the lengthiness of exposure to erosion, the level of slope, and land management most especially around loosely tilled soil and cover soil. Furthermore, it has been observed that there is tendency for the level of soil erodibility to increases when the level of soil organic matter drops under 2% [27, 47].

12.4.6.3 Nutrient Cycling

The level of soil organic matter has been identified could be linked to the amount of nutrient available in different soil [24]. It has been observed that these values may vary between different soils, pastures, history of crops, and management of land [59]. It has been observed that soil organic matter could be liberated into the soil which also enables the releases of these nutrients to plants however, this depends on several factors: during the process of decomposition of soil organic matter, recycling of nutrients, and movement of nutrients.

This shows that soil organic matter in a significant factor which shows that source of nutrients which constitutes an important factor in agriculture for effective increase in agricultural production and enhance in food production.

Nitrogen has been recognized as an essential nutrient for several biological activities. It has been observed that soil microorganism performs essential role in the biotransformation of nitrogen most especially among these various aforementioned pools. The beneficial microorganism possesses that capability to convert these soil organic matters through the process of mineralization by converting the nitrogen available in the soil organic matter, soil organisms, and plant materials to mineral nitrogen which exist as ammonium and nitrate ions which are liberated and release to plants inform of nitrogen. The level of efficiency of mineralization might be linked to the chemical characterization of the substrate minerals most especially its lignin content and C:N ratio.

Phosphorus has been observed to exist in numerous pools majorly in the soil and numerous phosphorous derived from soil organic matter most especially from different soil [49]. It has been showed that phosphorous derived from the soil organic matter is approximately 40% of the total phosphorus available when tested the soil are tested which might varies from 20% to 80% irrespective of the soil type.

Also, the presence of aluminum and iron might prevent the fixing of phosphorus into unavailable form to plant but they need to become mineralized before they could be made available to plants [33]. It has been observed that level of soil organic matter and mineralization of plants are the major source of phosphorus to plants [29].

Sulfur could be liberated from the total amount of soil organic matter available in most soil although some of the soil has been identified to possess a higher amount of sulfur most especially in some volcanic and gypsum while some of them are also present in marine deposits which area related to acidic sulphate soils. The presence of soil microbial population has been identified as necessary factor that are essential in soil organic matter. The maintenance of the soil health has been identified as critical factor that ensure the sustainability of the normal ecosystem of the soil and most especially the soil organic matter [60]. The management of land has a greater effect on the nutrient made available by soil beneficial microorganisms for easily absorption by plants for sustainable agriculture and increase in food production [32].

12.5 Conclusion and Future Recommendation to Knowledge

This chapter has highlighted the significant of numerous factors that are necessary for normal functioning of the soil factors. Special emphasis was laid on soil organic carbon as an essential part of the soil which also ensures the proper functioning of the ecosystem. It has been discovered that the accumulation soil organic carbon could be linked to several interaction among diverse dynamic ecological processes which includes soil respiration, decomposition, and respiration. Therefore, there is a need to strategically look into several human activities that could led to decrease in the level of soil organic carbon, global climate changes while there is a need to encourage human activities that will support numerous human actions that could sequestering carbon back into soil. There is a need to take into consideration some other relevant factors such as elevated CO_2. Future warming, land management strategies, patterns of past land use, along with the physical heterogeneity of landscapes should be taken into consideration. Moreover, this chapter has highlighted several merits and significant of pH as a pointer of soil biogeochemical processes in environmental research. Therefore, there is a need to highlight numerous biogeochemical processes that could be affected by soil pH and the relevance of their association for future planning, research, and development. Furthermore, detailed information on microbial ecophysiological indicators which could serves as a interlinkage that exist between the functioning of cell-physiological when subjected to the effect of numerous environmental factors were also highlighted. The utilization of the metabolic quotient (qCO2) as an index which normally indicate the effectiveness of organic substrate usage mainly through the help of beneficial soil microorganism in specific soil conditions were also highlighted. Also, significance of soil enzyme activities and their relationship with soil pH, biodegradation of toxic substances by soil microorganisms and their relationship with soil pH, water holding capacity, soil erodibility, the influence of beneficial microorganism, and the effect of soil enzyme on nutrient cycling are also discussed. There is a need for government, policy maker, and scientist to develop a robust policy that will support the several human practices that will preserve the maintenance of numerous soil functions. This will go a long way in maintenance of the soil structure, provision of adequate nutrients which could increase the growth of agricultural, thereby providing adequate and quality food for ever increasing population.

References

1. Ananbeh, H., Stojanović, M., Pompeiano, A., Voběrková, S., Trasar-Cepeda, C., Use of soil enzyme activities to assess the recovery of soil functions in abandoned coppice forest systems. *Sci. Total Environ.*, 694, 133692, 2019.

2. Anderson, T.-H., Microbial eco-physiological indicators to asses soil quality. *Agric. Ecosyst. Environ.*, 98, 285–293, 2003.
3. Anderson, T.-H. and Domsch, K.H., Carbon link between microbial biomass and soil organic matter, in: *Perspectives in Microbial Ecology*, F. Megusar and M. Gantar (Eds.), pp. 467–471, Mladinska Knjiga, Ljubljana, Slovenia, 1986.
4. Allison, V.J., Condron, L.M., Peltzer, D.A., Richardson, S.J., Turner, B.L., Changes in enzyme activities and soil microbial community composition along carbon and nutrient gradients at the Franz Josef chronosequence, New Zealand. *Soil Biol. Biochem.*, 39, 7, 1770–1781, 2007.
5. Arora, S., Arora, S., Sahni, D., Pesticides use and its effect on soil bacterial and fungal populations, microbial biomass carbon and enzymatic activity. *Curr. Sci.*, 116, 4, 643–649, 2019.
6. Blagodatskaya, E.V. and Anderson, T.-H., Interactive effects of pH and substrate quality on the fungal-to-bacterial ratio and qCO2 of microbial communities in forest soils. *Soil Biol. Biochem.*, 30, 1269–1274, 1998.
7. Bonanomi, G. *et al.*, Soil Biology and Biochemistry Organic farming induces changes in soil microbiota that affect agro- ecosystem functions. *Soil Biol. Biochem.*, 103, 327–336, 2016.
8. Brady, N.C. and Weil, R.R., *Nature and Property of Soils*, Prentice Hall, Upper Saddle Hall, NJ, USA, 1999.
9. Cullen, J.J. and Boyd, P.W., Predicting and verifying the intended and unintended consequences of large-scale ocean iron fertilization. *Mar. Ecol. Prog. Ser.*, 364, 295–301, 2008.
10. Dahlgren, R.A., Biogeochemical processes in soils and ecosystems: from landscape to molecular scale. *J. Geochem. Explor.*, 88, 186–189, 2006.
11. Dass, S., Muneer, M., Gopidas, K.R., Photocatalytic degradation of wastewater pollutants, Titanium-dioxidemediated oxidation of polynuclear aromatic hydrocarbons. *J. Photochem. Photobiol. A: Chem.*, 77, 83–88, 1994.
12. Drake, B.G., Gonzalez-Meler, M.A., Long, S.P., More efficient plants: A consequence of rising atmospheric CO_2? *Annu. Rev. Plant Physiol.*, 48, 608–637, 1997
13. Ehrenfeld, J.G., Kaldor, E., Parmelee, R.W., Vertical distribution of roots along a soil toposequence in the New Jersey pinelands. *Can. J. For. Res.*, 22, 1929–1936, 1992.
14. FAO and Revised World Soil Charter, Food and Agriculture Organization, Rome, Italy, FAO and ITPS, *Status of the World's Soil Resources (SWSR)—Main Report*, Food and Agriculture Organization of the United Nations and Intergovernmental Technical Panel on Soils, Rome, Italy, 2015.
15. Fierer, N. and Jackson, R.B., Diversity and biogeography of soil bacterial communities. *Proc. Natl. Acad. Sci.*, 103, 3, 626–631, 2006.
16. Gil-Sotres, F., Trasar-Cepeda, C., Leirós, M.C., Seoane, S., Different approaches to evaluating soil quality using biochemical properties. *Soil Biol. Biochem.*, 37, 877–887, https://doi.org/10.1016/J.SOILBIO.2004.10.003, 2006.

17. Gomez, E., Ferreras, L., Lovotti, L., Fernandez, E., Impact of glyphosate on microbial biomass and metabolic activity in a *Vertic argiudoll* from Argentina. *Eur. J. Soil Biol.*, 45, 2, 163–167, 2009.
18. Houghton, R.A., Balancing the global carbon budget. *Annu. Rev. Earth Planet. Sci.*, 35, 313–347, 2007.
19. Hungate, B.A. *et al.*, The fate of carbon in grasslands under carbon dioxide enrichment. *Nature*, 388, 576–579, 1997.
20. Intergovernmental Panel on Climate Change (I.P.C.C), *Climate Change 2007: The Physical Science Basis*, Cambridge University Press, Cambridge, UK, 2007.
21. Jones, A., Breuning-Madsen, H., Brossard, M., *Soil Atlas of Africa*, European Commission, Publications Office of the European Union, Brussels, Belgium, 2013.
22. Jung, J., Philippot, L., Park, W., Metagenomic and functional analyses of the consequences of reduction of bacterial diversity on soil functions and bioremediation in diesel-contaminated microcosms. Nature Publishing Group. 6, 10pp, 2016.
23. Keeler, S.L., Sinsabaugh, R.L., Crenshaw, C., Pulse dynamics and microbial processes in arid land ecosystems. *J. Ecol.*, 96, 3, 413–420, 2008.
24. Kirkby, C.A., Kirkegaard, A., Richardson, A.E., Wade, L.J., Blanchard, C., Batten, G., Stable soil organic matter: A comparison of C:N:P:S ratios in Australian and other world soils. *Geoderma*, 163, 197–208, 2011.
25. Lal, R., Soil carbon sequestration impact on global climate change and food security. *Science*, 304, 1623–1627, 2004.
26. Lemus, R. and Lal, R., Bioenergy Crops and carbon sequestration. *Crit. Rev. Plant Sci.*, 24, 1, 1–21, 2005.
27. Loch, R.J. and Foley, J.L., Measurement of aggregate breakdown under rain: Comparison with tests of water stability and relationship with field measurement of infiltration. *Aust. J. Soil Res.*, 32, 4, 701–720, 2004.
28. Maracchi, G., Sirotenko, O., Bindi, M., Impacts of present and future climate variability on agriculture and forestry in the temperate regions: Europe. *Clim. Change*, 70, 117–135, 2005.
29. McLaughlin, M.J., McBeath, T.M., Smernik, R., Stacey, S.P., Ajiboye, B., Guppy, C., The chemical nature of phosphorus accumulation in agricultural soils – implications for fertilizer management and design: an Australian perspective. *Plant Soil*, 349, 69–87, 2011.
30. McNeill, J.R. and Winiwarter, V., Breaking the sod: humankind, history, and soil. *Science*, 304, 1627–1629, 2004.
31. Millennium Ecosystem Assessment, *Ecosystems and Human Well-Being: Desertification Synthesis*, World Resources Institute, Washington, DC, USA, 2005.
32. Minasny, B., Hong, S.Y., Hartemink, A.E., Kim, Y.H., Kang, S.S., Soil pH increase under paddy in South Korea between 2000 and 2012. *Agric. Ecosyst. Environ.*, 221, 205–213, 2016.

33. Moody, P.W. and Bolland, M.D.A., Phosphorus, in: *Soil Analysis - an interpretation manual*, K.I. Peverill, L.A. Sparrow, D.J. Reuter, (Eds.), CSIRO Publishing Collingwood, Australia, 2005.
34. Mukherjee, S., Tripathi, S., Kumar, A., Persistence of the herbicides fl orasulam and halauxifen-methyl in alluvial and saline alluvial soils and their effects on microbial indicators of soil quality. *Eur. J. Soil Biol.*, 73, 93–99, https://doi.org/10.1016/j.ejsobi.2016.01.009, 2016.
35. Neina, D., The Role of Soil pH in Plant Nutrition and Soil Remediation. *Appl. Environ. Soil Sci.*, 2019, 9, 2019, Article ID 5794869.
36. Neina, D., Buerkert, A., Joergensen, R.G., Microbial response to the restoration of a Technosol amended with local organic materials. *Soil Tillage Res.*, 163, 214–224, 2016.
37. Neina, D., Buerkert, A., Joergensen, R.G., Effects of land use on microbial indices in tantalite mine soils, Western Rwanda. *Land Degrad. Dev.*, 28, 1, 181–188, 2017.
38. Ontl, T.A. and Schulte, L.A., Soil Carbon Storage. *Nature Education Knowledge*, 3, 10, 35, 2012.
39. Oelkers, E.H. and Cole, D.R., Carbon dioxide sequestration: a solution to the global problem. *Elements*, 4, 305–310, 2008.
40. Ostle, N.J., Levy, P.E., Evans, C.D., Smith, P., UK land use and soil carbon sequestration. *Land Use Policy*, 26, S274–SS283, 2009.
41. Pataki, D.E. *et al.*, Tracing changes in ecosystem function under elevated carbon dioxide conditions. *BioScience*, 53, 805–818, 2003.
42. Pawar, R.M., Effect of soil pH on bioremediation of polycyclic aromatic hydrocarbons (PAHS). *Bioremediat. Biodegrad.*, 6, 3, 1–14, 2015.
43. Pietri, J.C.A. and Brookes, P.C., Nitrogen mineralisationalong a pH gradient of a silty loam UK soil. *Soil Biol. Biochem.*, 40, 3, 797–802, 2008.
44. Pietri, J.C.A. and Brookes, P.C., Relationships between soil pH and microbial properties in a UK arable soil. *Soil Biol. Biochem.*, 40, 7, 1856–1861, 20082008.
45. Rawls, W.J., Ahuja, L.R., Brakensiek, D.L., Estimating soil hydraulic properties from soils data, in: *Indirect methods for estimating hydraulic properties of unsaturated soils. Proceedings of International Workshop on Indirect Methods for Estimating the Hydraulic Properties of Unsaturated Soils. Riverside, California*, October 11–13, 1989 US salinity Laboratory, Agricultural research Service, US Department of Agriculture, Riverside, California.
46. Riah, W., Laval, K., Laroche-Ajzenberg, E., Mougin, C., Latour, X., Trinsoutrot- Gattin, I., Effects of pesticides on soil enzymes: a review. *Environ. Chem. Lett.*, 12, 257e273, 2014.
47. Rosewell, C.J. and Loch, R.J., Estimation of the RUSLE soil erodibility factor, in: *Soil Physical Measurement and Interpretation for Land Evaluation*, N. McKenzie, K. Coughlan, H. Cresswell (Eds.), CSIRO Publishing, Collingwood, Australia, 2002.

48. Sánchez-Rodríguez, A.R., Hill, P.W., Chadwick, D.R., Jones, D.L., Typology of extreme flood event leads to differential impacts on soil functioning. *Soil Biol. Biochem.*, 129, 153–168, https://doi.org/10.1016/j.soilbio.2018.11.019, 2019.
49. Shen, J., Yuan, L., Zhang, J., Li, H., Bai, Z., Chen, X., Zhang, W., Zhang, F., Phosphorus dynamics: from Soil to Plant. *Plant Physiol.*, 156, 997–1005, 2009.
50. Sinsabaugh, R.L., Carreiro, M.M., Alvarez, S., Enzyme and microbial dynamics during litter decomposition, in: *Enzymes in the Environment: Activity, Ecology, and Applications*, R.G. Burns and R.P. Dick (Eds.), pp. 249–266, CRC Press, Inc., Oxford, UK, 2002.
51. Sinsabaugh, R.L., Lauber, C.L., Weintraub, M.N., Stoichiometry of soil enzyme activity at global scale. *Ecol. Lett.*, 11, 11, 1252–1264, 2008.
52. Stursova, B.K. and Walker, A., Microbial degradation of organophosphorus compounds. *FEMS Microbiol. Rev.*, 30, 3, 428–471, 2006.
53. Thompson, J.A. and Kolka, R.K., Soil carbon storage estimation in a central hardwood forest watershed using quantitative soil-landscape modeling. *Soil Sci. Soc. Am. J.*, 69, 1086–1093, 2005.
54. Turner, B.L., Variation in pH optima of hydrolytic enzyme activities in tropical rain forest soils. *Appl. Environ. Microbiol.*, 76, 19, 6485–6493, 2010.
55. Vidali, M., Bioremediation. An overview. *Pure Appl. Chem.*, 73, 7, 1163–1172, 2001.
56. Voos, G. and Groffman, P.M., Relationship between microbial biomass and dissipation of 2,4-D and Dicamba in soil. *Biol. Fertil. Soils*, 24, 106–110, 1997.
57. Wang, F., Li, X., Zhu, L., Du, Z., Zhang, C., Wang, J., Wang, J., Lv, D., Responses of Soil Microorganisms and Enzymatic Activities to Azoxystrobin in Cambisol. *Pol. J. Environ. Stud.*, 27, 6, 2775–2783, 2018.
58. World Meteorological Organization (W.M.O), *Greenhouse Gas Bulletin*, World Meterological Organization, Geneva, Switzerland, 2006.
59. Williams, J., Ross, P., Bristow, K., Prediction of the Campbell water retention function from texture, structure and organic matter, in: *Indirect methods for estimating hydraulic properties of unsaturated soils. Proceedings of International Workshop on Indirect Methods for Estimating the Hydraulic Properties of Unsaturated Soils*, Riverside, California, October 11–13, 1989 US salinity Laboratory, Agricultural research Service, US Department of Agriculture, Riverside, California.
60. White, R.E., *Principles and Practice of Soil Science: Soil as a Natural Resource*, Blackwell Publishing, Oxford, UK, 2006, https://doi.org/10.1155/2019/5794869.
61. Williams, C.H. and Donald, C.M., Changes in organic matter and pH in a podzolic soil as influenced by subterranean clover and superphosphate. *Aust. J. Agric. Res.*, 8, 179–89, 2005.

62. Xiong, X., Grunwald, S., Myers, D.B., Ross, C.W., Harris, W.G., Comerford, N.B., Interaction effects of climate and land use/land cover change on soil organic carbon sequestration. *Sci. Total Environ.*, 493, 974–982, 2014.
63. Yao, X., Min, H., Lü, Z., Yuan, H., Influence of acetamiprid on soil enzymatic activities and respiration. *Eur. J. Soil Biol.*, 42, 120–126, https://doi.org/10.1016/j.ejsobi.2005.12.001, 2006.
64. Zak, D.R. *et al.*, Elevated atmospheric CO_2, fine roots and the response of soil microorganisms: A review and hypothesis. *New Phytol.*, 147, 201–222, 2000.

13

Role of Pesticide Applications in Sustainable Agriculture

Osikemekha Anthony Anani[1] and Charles Oluwaseun Adetunji[2]*

[1]Laboratory of Ecotoxicology and Forensic Biology, Department of Biological Science, Faculty of Science, Edo University, Iyamho, Edo State, Nigeria
[2]Applied Microbiology, Biotechnology and Nanotechnology Laboratory, Department of Microbiology, Edo University Iyamho, Auchi, Edo State, Nigeria

Abstract
The introduction of industrialization has led to several revolutions in the area of agricultural sector which has enhanced the amount of chemical burden on natural ecosystems. It has been discovered that pesticides are agrochemicals that are utilized in urban areas, agricultural lands and public health programs so as to protect humans and plants from numerous diseases. Moreover, it has been discovered that there are several detrimental such as environmental and health effect which makes their influence to be a crucial environmental health risk factor. Therefore, this chapter intends to provide detailed information on the application of pesticides in agriculture. The merits and demerits of pesticides were also highlighted. The modes of action through which most of these pesticides are applied were also stated.

Keywords: Agriculture, environmental, health, modes of action, industrialization, agrochemicals, ecosystems

13.1 Introduction

Agriculture has been identified to play a crucial function toward the achievement of sustainable development and reduction of poverty majorly in the middle-income and low-income countries [1]. Agricultural pest and

**Corresponding author*: charliguitar@yahoo.com; adetunjicharles@gmail.com; adetunji.charles@edouniversity.edu.ng

Inamuddin, Mohd Imran Ahamed, Rajender Boddula and Tariq Altalhi (eds.) Applied Soil Chemistry, (235–256) © 2021 Scrivener Publishing LLC

diseases have been recognized as a major factor that affects increase in agricultural development because these cause drastic reduction in agricultural productivity and cause diseases in crops [2]. Numerous technologies have been developed to minimize the incidence of pests but the utilization of pesticides has been discovered to mitigate against all the aforementioned challenges facing increase in agricultural productivity [3].

Agricultural pesticides entail numerous chemicals majorly applied by farmers to prevent and destroy the influence of pests. These chemicals are very active in minimizing agricultural pests; they have also been recognized to have a detrimental effect on human health, animal, and ecosystem [4, 5]. Pesticides are synthetic substances or the combination of constituents that are majorly utilized in public health protection programs or in sustainable agriculture against pests and diseases, weeds, as well as their application against human vector-borne diseases such as schistosomiasis, malaria, and dengue fever, respectively.

Some examples of these pesticides include herbicides, plant growth regulators, insecticides, rodenticides, and fungicides [6–8].

The application of these pesticidal products is utilized for the enhancement and improvement of some non-agricultural activities such as sport fields and public areas in urban environment [9, 10]. Hence, some other application of these pesticides includes building materials, pet shampoos, and bottom part of ships so as to prevent or minimize the occurrence of undesirable species [11]. The usage of these pesticides has been related to environmental and health issues [12–17], which has led to the unrestricted application of some of these pesticides. It has been observed that these pesticides may be metabolize, excreted, storage, or building up in human body [18].

Some of the detrimental health influence includes endocrine effects, dermatological, reproductive, gastrointestinal, carcinogenic, and neurological [18–23]. Also, intentional, high occupation, and accidental led to high level of mortality and hospitalization [24]. Moreover, there is a tendency for a high level of pesticides residue in post-harvest products, fresh products, foods, and beverages [25, 26]. In addition, the presence of pesticide residue has been recorded by some scientists in human breast milk samples [27–29].

Therefore, this chapter intends to provide detailed information on the application of pesticides in agriculture. This chapter also elaborated on the numerous types of pesticides which include neonicotinoids, pyrethroids, organochlorines, organophosphates, triazines, carbamates, and pyrethroids. The merits and demerits that are associated with chemical pesticides were also highlighted. The modes of action and the detrimental effects of these pesticides were also stated.

13.2 Various Types of Pesticides and Their Application in Agriculture

The management of agricultural ecosystem from pest invasion is very essential to plant's and animal's productivities. However, the application of noxious substances to destroy pests has an extent proven to be important asset in agricultural management. In view of this, Lichtenberg, [30] did a review of the application, utilization, and economics of pesticides in agricultural purpose. The authors stated that pesticides' application in agricultural activities is important exploitation to compliment agricultural processes in order to reduce the impact caused by natural foes called pests. That the management of the agricultural ecosystem requires special resistance management pesticide delivery system in the precise improvement of crops and to combat adverse impacts of pesticides misapplications. However, of recent, there have been several types of regulatory attempts to reduce the human and environmental influence of pesticides by limiting the condition and availability of usage. They stated that the use of small quantity in the farm consistently over a period of time can have an overall significance influence and distributional impacts on the ecosystem, floras, and faunas. In conclusion, the authors stated that extensive precincts on pesticides application and use should be put in place in the management of agricultural produce. The safety of farmworkers should also be put into considerations as well as the end consumers along the food chain in order to avoid bioaccumulation and toxicity to humans.

Watanabe *et al.* [31], in a book chapter review, looked at application of pesticides as well as the exposure evaluation in rice paddy regions of Japan. They also looked at the modeling methods and the monitoring of pesticides in ground and surface water systems. They stated that Japan is one of the countries of the world that exploit the use of pesticides in agricultural activities specifically for rice production as compared to the USA and EU. They stated that for sustainable and eco-friendly production of rice by utilizing pesticides, an up-to-date development and methods in evaluating climatic conditions and pesticides pollution are inevitable. They reported that most rice paddies are treated with insecticides, fungicides, and herbicides for combating insects, fungi, and killing weeds. The application of novel fungicides and insecticides is also used during seed-bed groundwork. In conclusion, they recommend an overall assessment of the possible health and environmental impacts pesticides that portend the agroecosystem.

The population of the world is increasing day in day out globally. There is the need to increase the output of crop production in order to feed the ever teeming population. However, the use of chemicals such as fertilizers and pesticides in agricultural activities are very diverse in the boosting of plant production. On this note, Cheremisin, [32] did a review of the best way to apply pesticides for a cleaner crop production and pollution control. They opined that there are five best practices to meet the quality agriculture demands for crop production which are 1) reducing or controlling fertilizer application, 2) regulating pests by biological and chemical techniques, 3) by expanding and introducing irrigation method, 4) by improving plant varieties, and 5) improving land tillage techniques. The first two are known to have eco-benefits to large-scale production as well as in the protection of plants against pests and diseases and for sustainable ecosystem to human health.

Population explosion has definitely affected food production. The demand for food as a result of population growth is *sine qua non* to high standards of living of recent. Nonetheless, there is the need to boost crop production. In line with this, Liu *et al.* [33], in a review, looked at the application and use of pesticides and fertilizers on cereal production. The authors stated that minerals from fertilizers and pesticides can be used to improve crop production. However, pesticides can impact negatively to the environment as well as food crops, thus resulting to world-wide food crisis. Findings from their review study showed that countries indicating reduction or fluctuation yields have GDP (gross domestic product)/capital, specifically in West Asia, South America, and sub-Saharan African, that the yields (47%) and production (38%) of cereal of countries with low and floating GDP decreased from 1961 to 2010 and pesticides and fertilizer consumption of the low GDP countries rise from 35% to 40%. The cause and effects of this upturn of production, yields, and GDP are evidences of imbalance in the global use of pesticides and fertilizers as it relates to food production. Therefore, there the need to optimize the consumption of chemical pesticides and fertilizers for more economic advances, environmental protection and landscape production with extensive control, and regulation and management to forestall a sustainable agro-system and environment.

To meet the growing need of food security in the ever increasing population, several methods need to be applied for the enhancement of crop production in order to meet the SDGs (sustainable development goals). Rosenfeld and Feng, [34] did a review of the potential risk in the utilization of pesticides and fertilizers. They stated that using pesticides and fertilizers is strictly to control pest and diseases as well to boost crop production. The increase in their productions, have pave way for many agro-industry to strive. That along the positive benefits they portend are the snowballing influence on the environment resulting to varied human health and environmental assets damages and decorations. They reported that pesticides pollution to drinking, surface, and ground water has come to be a very universe problem to many developed countries of the world like USA. He evaluated five pesticides that usually found in consume water, where two of these, DBCP (1,2-dibromo-3-chloropropane) and alachlor, are categorized as potential carcinogens to humans linked to rise in organ damage and malignant tumour in animals. The rebound of pests and the upsurge of pesticides application have become the fragment of the equivalence challenges which require a balance of the overall health of the ecosystem. In conclusion, the authors suggested strict rules and arduous testing to be adopted by farmers and relevant stallholders in the regulation and management of pesticides application in order not to pose both environmental and health risks to the ecosystem and the food chain at large.

Sánchez-Bayo, [35], in a review, looked at the influence of pesticides on agroecosystems as well as possible abatement and risk management strategy. The author recounted that pesticides are noxious chemicals use in the regulation of pathogens, weeds, and pests. That about ¾ of the total pesticides is use in agricultural activities specifically in developed nations in the quest to reduce crop injury. Nonetheless, after about 60 years of their utilization, their influence in the terrestrial ecosystem still remains noticed. The author stated the impacts of fungicides, herbicides, insecticides, and pesticides on the soil, plants, and animals. That fungicides change the consortia of fungi-microbes by reutilizing the soil nutrients, while Cu fungicides are noxious to many animals especially earthworms. The continuous use of herbicides on weeds has amounted to negative loss of biodiversity and biomass in many land settings as well as the circuitously decreases of the link between population drop of amphibians, birds, and arthropods with food insecurity. The toxic influence of insecticides on sol invertebrate population, aves, and mammals has resulted to rapid decline in their distribution and population, as well as the community structure and trophic level of food and energy transfer. The persistent accumulation of pesticides in the soil has increased in terms of concentration. The authors stated that the trophic food chain structure has been affected by plethora of sub-fatal negative influence of pesticides. That the long-term impacts of cyclodiene and DDT poisoning in aves still remain in the ecosystem despite the stoppage of it for the past 30 years in most developed nations. In conclusion, the author recounted that usage of pesticides in agricultural activities has really helped to boost crop yield and production. However, their impacts on the ecosystem are very great.

Any organism that results to physical and chemical damage to agricultural produces thus reducing their economic worth is termed farm pest. However, chemicals are used to repel or exterminate pest in order to prevent the destruction of crops and diseases to plants and humans. In this context, Pandya, [36] did a review of the application of pesticides in agriculture. The author recounted that most pesticides are have the potential to be organism-specific and have specific mechanism of action to eliminate pests. There is no doubt that pesticides enhance the productivity of crop production and promote growth and development as well as the economy output. However, the negative influence on the biotic and abiotic components of the ecosystem is striking. The quality of the air, water, and soil is compromised via biomagnification. Majority of pesticides are carcinogenic and mutagenic to humans and plants. The author stated that there is a clarion call to produce environmental friendly pesticides that can reduce the negative impacts faced from its use. He recommended, in conclusion, the use of eco-friendly natural pesticides such as Chile pepper, *Chrysanthemum* tea flower, garlic, onion, oils from Eucalyptus, mineral, citrus, and Neem plants for combating pests and diseases of crops. That tobacco plants were formerly used as pesticides for exterminating aphids, caterpillars, and related farm pests, however, proven to be carcinogenic and hazardous to humans.

Maton *et al.* [37], in a review, looked at the effects of pesticides utilization in agricultural farms in Nigeria. The authors recounted the aftermath problem encounter in the quest to control pest infestation in agricultural land by using pesticides. Because of the effectiveness, affordability, and availability of pesticides, such as herbicides, insecticides, and fungicides, the persistence utilization in farm activities is alarming causing impounding rebound impacts to water, soil, floras, and faunas. The authors, in conclusion, recommended that public awareness and judicial measures should be put in place to regulate the uncontrolled use of pesticides. The return to crude methods of farming weed with hand and hoes, will aid in forecasting future degradation of the ecosystem, hence encouraging sustainable farming for the present and future generation.

In modern day agriculture, pesticides are used to regulate weeds and pests as well as pre- and post-harvest protection of plant and crops. However, the ecological and health impacts on humans cannot be overemphasized. Mattah *et al.* [38], in a review, looked at the health implication in the application of pesticides during farming activities in Ghana. One hundred and twenty farmers who utilize, handle, and acquire pesticides during farming activities as well as 12 brands of various pesticides, such as fungicides, insecticides, and herbicides, were examined. From their outcome, three major points (10% from farmers, 39.2% from agricultural chemicals traders, and 43.3% from extension workers) of information were acknowledged. However, it was observed that 75% of those who responded bought pesticides from agricultural chemicals factories. In total, 25.7% of the 74 farmers were noticed to sprinkle pesticides on the farmlands with full safety kits, while 66.7% of the farmers, whose pesticides got finished, abandoned the materials of spraying on their farmlands or dispose them in nearby bushes. It was also observed that the rate of application of the pesticides was subjected by the size of the farmland and the inexpensive nature of the chemicals. In conclusion, the farmers recommend the training of stakeholders and farmers on how to apply and handle pesticides as well the proportion to use depending on the intensity of the pest invasion or rebound.

13.3 Modes of Action of Pesticides

Pesticides are well known for their recalcitrant and persistence nature in the ecosystem. The biotransformation ability of pesticides enables them to remain in living and non-living things in the environment. Lushchak *et al.* [39], in a review, looked at the mechanism of action of the toxic nature of pesticides. The authors classified pesticides based on their pathophysiological impacts, physical nature, use, sources, and their intrinsic nature. The impact of these environmental xenobiotics can be deciphered based on their molecular actions. Pesticides such as insecticides, fungicides, and herbicides have specific modes of actions when applied to crops. They stated that insecticides are known for their influence on energy production, development and growth of plants, and impact the muscles and nerves of humans. Fungicides are strictly inhibitors of mitochondrodrial-respiration activities, protein site synthesis, and ergo-sterol synthesis, whereas herbicides are basically regulators disrupters of pigment synthesis, cell membrane, lipid and amino acid synthesis, and growth and development of seeds and roots of plants. The authors recounted that the awareness of the potential health risks to human is very paramount in order to avoid possible diseases like nerve damage, disease invasion, inflammation, atherosclerosis, malnutrition, and unstable metabolic condition. In conclusion, the authors recommend the study of the long term impact of pesticides and dosages on non-target species. More so, a novel eco-friendly technology should be employed in the farming of crops in order to avoid ecosystem contamination and possible risks to human.

Rajashekar and Shivanandappa, [40] evaluated the mechanism of action of decaleside-pesticides. The authors recounted that a novel type of pesticides decaleside are known for their noxious impact to arthropods. That this novel pesticides attack the tarsal gustatory chemosensilla of the arthropods. The rapid breakdown and toxicity symptoms of the pesticides proposed neurotoxic mechanistic action. However, specific molecular-targets and mode of actions, currently, are not known. The authors also stated that in their previous study, they reported the inhibitory pesticidal actions of decaleside in the connection with Na pump in housefly and cockroach. This was also similar to what they obtained from an *in vivo* inhibition study of –ATPase, K+, and Na+ in the thorax and head of arthropods, blocking and ablation of tarsi locations. A trisaccharide, maltotriose, aids in the temporary liberation of the noxious state of decaleside as well as the hindrance of the organic catalyse (enzyme) which signify the probable contribution of sugar gustatory receptors. In addition, an *in vitro* experimental studies with simple enzymes sourced, purified, and prepared from porcine –ATPase, K+, and Na+ revealed that the pesticide decaleside reduced the organic catalyse via competing with the ATP bond-site of respiration. In conclusion, the authors stated that the mechanism of action decaleside through tarsal gustatory locations can be associated to the reduction of Na pump which indicate an exclusive mechanistic action. That further studies need to be carried on decaleside in order to ascertain the pathway of tarsal chemosensilla VS the Na pump pesticidal actions.

Casida, [41], in a review, looked at the mechanisms of action of pesticides. The authors stated that approximately 700 pesticides, such as fungicides, herbicides, and insecticides, act on 95 biochemical bull's eye in the annihilation of weeds, fungi, insects, and pests. He stated that, of present, insecticides basically act on four nerve bull's eye which are the γ (gamma)-aminobutyric acid receptors, acetylcholine receptors, the voltage-gated Cl network, and the acetylcholinesterase, which are not present in plants but animals. The mechanism of action of herbicides is via impeding the pathways of amino acid, carotenoid, and photo synthesis that are vital to plants not animals. Some fungicides impede the biosynthesis of reductase cytochrome-c and ergosterol, whereas some upset primary cellular activities. The author stated that pesticides show a range of limiting factors, that is, acting on the same organisms, not resilient to specific chemical components and been selective on strains of organisms. In conclusion, the author recommends a persistent pesticide and pest management strategy by understating the relativity of molecular and biochemical toxicology of crops, humans, and pests via application of ideologies of human health food production and pest physiology.

Casida, [42], in a review, evaluated the metabolism and modes of action of insecticides, organo-phosphorus, as it relates to their left over properties, stability, and activity. The author recounted that organo-phosphorus are active on living organisms by inhibiting the organic catalyst via the production of an inactive form known as phosphoryl esterases. In some cases, stable forms of it can be converted in some cases in the living organism to more responsive antiesterases. However, not all, some phosphorothioates and dimethylphosphoramides need an *in vivo* radical oxidation state to impact on the action of the antiesterase. In addition, some biochemical activities either reduce or increase the level of organo-phosphate antiesterase actions. For example, the importance of phosphoryl esterase cannot be overemphasis in the stability of the process of enzymes unstableness and might aid in detoxification of contaminants in the cell. In another phase of very stable *in vivo* reaction, phosphoryl cholinesterase may result to chronic or sub-lethal toxicity condition. In conclusion, the author stated that an excellent organo-phosphate needs a very a good balance of stability, antiesterase action, and specificity for mechanistic functioning on any substrate.

United States Environmental Protection Agency, [43] did a review of the extensive application as well as the modes of actions of some insecticides used in the regulation of insects. The environmental regulation bodies specifically stated that most of these insecticides target the nervous system and inhibit the enzymes; cholinesterase thereby regulating growth in the organism. That organochlorine attacks the neurone by resulting to Na-K acid and γ-aminobutyric imbalance thus impeding nerve impulse transmission and hyperexcitable condition (convulsions and tremors). Organophosphate results to the inhibition of acetylcholinesterase as well as in the build-up of acetylcholine at the junction of neuromuscular. This causes paralysis and rapid muscular twitching in vertebrates. Organosulfur displays ovicidal action by causing the death of eggs at every stage of development inmates as well as related species. Carbamates result to the inhibition of acetylcholinesterase, resulting to CNS paralysis and quick twitching of conditional muscles in fish. Formamidines reduces the monoamine oxidase, an enzyme that breaks down neurotransmitters, resulting to bioaccumulation of the pesticides thus affecting pests and insects-death. Dinitrophenols function by inhibiting and uncoupling the phosphorylation oxidative cycles, thus impeding the production of ATP. However, there is a ban on its general usage. Organotins reduce the phosphorylation location of dinitrophenol forming, preventing and uncoupling the ATP site of actions against fruit-tress and mites. In addition, it's is also use as molluscacide and antifouling messengers, which have proven noxious to aquatic biota. Pyrethroids are well known for its protection of open Na channel in the neurone-membrane impacting on both the CNS and PNS, a hyper-excitable condition. Possible indications are paralysis, hyperactivity, incoordination, and tremors in agricultural pests, insects, and fish. Spinosyns act as a disruption of acetylcholine by binding with acetylcholine receptors during cell postsynaptic. Its impact is well shown in termites, thrips, leaf miners, lepidopteran larvae, and caterpillars. Neonicotinoids act on the CNS resulting on the permanent blockage of the acetylcholine-nicotinergic postsynaptic receptors. It is used basically in controlling Colorado potato beetle, turf insects, termites, whiteflies, soil insects, and sucking insects. However, it is considered toxic at low concentration aquatic biota. Pyrazoles reduces the mitochondrial- electron transportation of cell at the site of NADH-CoQ reductase, thus interrupting with the formation of ATP. It is more efficient against thrips, whitefly, aphids, and psylla. One of its components, fipronil, has been confirmed to have no impacts to fish, oyster, and clams but low toxicity on shrimps. Pyridazinones disturb the transportation of mitochondrial electrons. Quinazolines act in blocking and inhibiting the synthesis of chitin from the exoskeleton in arthropods at the larval stage, thus resulting to the rupturing of the cuticle and demise via starvation. Pyrethrum affects both the PNS and CNS, thus motivating the nerve cells to synthesis continuous discharges which can cause death specifically in lice. Nicotine copycats the acetylcholine in the CNS ganglia causing death, convulsions, and rapid muscle twitching in caterpillars and aphids. Rotenone acts as an inhibition to formation of enzymes of respiratory. Its utilization as piscicide kills fish as food for other organism at low noxious dosage. Limonene impacts the PNS and CNS of in ticks, mites, lice, and fleas. But, low toxicity has been recorded in aquatic biota. Neem decreases and disrupts moulting and feeding by impeding biosynthesis of ecdysone; a hormone in juvenile during moulting in the larvae butterfly and moth. Antibiotics act as a disrupter of neuron transmission, at the junction of neurotransmitter. Fumigants play a role as narcotics that house in the fatty tissues thus causing narcosis or unconsciousness in insects.

Sherwani *et al.* [19], in a book review, looked at the mechanism of action of various types of herbicides as well as their safety and physiology. The authors stated that hierarchy organization, classification, and management of herbicides give an insight on the general physiology and sustainable strategy of how they can be utilize in farm activities in the regulation of unwanted plants. That the wrong application of herbicides in the right proportion like insecticides a form of insecticides can results to destruction, injury to weeds, as well as resistance. They reported that the modes of action of herbicides is based on the chemical structure which they use in the regulation and disruption of the biochemical properties of plants development and growth that are susceptible and eventual death ensue. In conclusion, the authors said that in-depth awareness of herbicide's mechanism of action is paramount in the specific management of symptomatic injury it may portend.

13.3.1 Organochlorine Pesticides

These are recognized as organochlorine pesticide which is also known as dichlorodiphenyltrichloroethane such as insecticide DDT. Their application may results into human and environmental hazards when injected or when applied in an uncontrolled manner [7, 44, 45]. Examples of organochlorines-based pesticides include methoxychlor, endosulfan, dieldrin, dicofol, and heptachlor. It has been identified that dichlorodiphenyldichloroethylene possesses carcinogenic effect and endocrine-disrupting influence [44]. Also, dichlorodiphenyltrichloroethane and DDT could be related to neurodevelopmental influence in children [46]. The diverse effect of dichlorodiphenyldichloroethylene has been documented to induce hepatic lipid dysfunction in rats [46] while some organochlorine pesticides has been affirmed to be responsible for some adverse effect on the following biological processes such as embryonic development Tiemann, [47], hepatic and hematological changes Freire *et al.* [48] endocrine disorders Mnif *et al.* [15] and Lemaire *et al.* [49], lipid metabolism Karami-Mohajeri and Abdollahi, [50] and carcinogenic effect [51, 52].

13.3.2 Organophosphorus Pesticides

Organophosphates has been identified as a more enhanced ecological substitute to organochlorines which entails numerous arrays of pesticides such as dimethoate, malathion, parathion, and glyphosate which constitute the most common among all. They possess endocrine-disrupting capability [53]. It has been observed that this group of pesticides could inhibit the role of cholinesterase enzymes Jaga and Dharmani, [53], reduce the insulin secretion, and affect the process of normal cellular metabolism involved in carbohydrates, protein, and fat Freire *et al.* [48], influence due to genotoxic Li *et al.* [54] and they have a detrimental effects on mitochondrial function which could later result into endocrine, cellular oxidative stress, and nervous disorder. Moreover, the effects of organophosphorus pesticides has been affirmed to cause serious health effect such as heart related diseases Hung *et al.* [55], impairment of the reproductive organ of male Jamal *et al.* [56], dementia Lin *et al.* [57], nervous system Eskenazi *et al.* [58], reduced gestational period Eskenazi *et al.* [59], and challenges that are neurological in small progeny [60].

Glyoposate has been reported to entails numerous adverse effect most especially in agriculture [61–66], when applied as an herbicides but the application of glyphosate-tolerant genetically modified crops most especially from crops such as maize and soybean [67–70]. Moreover, the application of genetically modified soybean cultivation has triggered some synergistic estrogenic influence which might be continues contact with to glyphosate as well as phytoestrogen "genistein," which entails common isoflavone available in soybean products and soybeans [71, 72]. It has been observed that glyphosate possesses the capability to cause endocrine-disrupting action [71, 72], detrimental influence on human erythrocytes during *in vitro assay* Kwiatkowska *et al.* [73], and enhance when tested on mouse skin [74]. Also, glyposhate could result into high level of destabilization in shikimate pathway which constitutes a major pathway in bacteria and plants as well as in the gut of human bacteria. It has been documented that the high-rate carcinogenicity of destabilization may inhibit the supply of amino acids to human organism [75].

It has been reported that commercial formulation entails a higher level of toxicity when compared to the activity of active substance alone [73, 76]. It has been stated that roundup which is a typical example of Glyphosate-based herbicides possesses the capability to cause endocrine disruptors and destruction in rat testicular cells and human cell lines enhance apoptosis in the testicular cells of experimental animals [77]. Furthermore, it has been observed that this group of pesticides could also affect the process of intracellular transport and cytoskeleton [78]. It has been reported that glyphosate could result into various types of health challenges such as cancer, strokes, hypertension, Parkinson's and Alzheimer's diseases, autism, diabetes, and kidney failure [72], while glyphosate possesses related deficiency available in essential trace metals, enhanced tendency to develop non-Hodgkin's lymphoma, and reproductive issues [79].

13.3.3 Carbamate Pesticides

It has been established that endocrine-disrupting involves some group of chemicals which fell under Carbamate pesticides. Typical examples of these includes ziram, aldicarb, and carbofuran; some examples of the adverse effect associated with this include adverse effect on mitochondrial role and modes of action of cellular metabolic and numerous challenges in the reproductive organs [80]. It has been established that carbamate pesticides possess the capability to induce cytotoxic as well as genotoxic influence most especially in the hamster ovarian cells Soloneski *et al.* [81] necrosis and cell death most especially in the immune of human cells Li *et al.* [54], regular destroyer of cells [82, 83], and cell death most especially against the lymphocytes [84]. Also, carbaryl could be categorized under carbamate pesticides which could function as a ligand most especially for the hepatic aryl hydrocarbon receptor which constitute a typical example of transcription factor which play a crucial role on the [84]. Also, carbamate pesticides could induce neurobehavioral effect, non-Hodgkin's lymphoma, and enhanced risk for dementia [85, 86].

13.3.4 Other Classes of Chemical Pesticides

This group of pesticides has been shown to induce reproductive toxicity and endocrine-disrupting influence. Typical examples are simazine, ametryn, and atrazine. It has been affirmed that there is a statically relationship between breast cancer incidence and triazine herbicides. Some of the adverse effect includes Atrazine, which has been recognized as a known type of triazines, that could be applied as herbicides which entail several adverse effects such as delays in sexual maturation, dopaminergic effects, oxidative stress, reproductive toxicity, and cytotoxicity [57, 87–91].

Some examples of synthetic pyrethroids include sumithrin, fenvalerate, and permethrin, which are known as safer insecticides which are made available for public health and agricultural purposes [92, 93]. They also possess the capability to show endocrine-disrupting action as well as possess the potential to affect reproductive parameters such as reproduction behavior most especially in experimental animals. It has also been stated that they possess the capability to show developmental neurotoxicity [94, 95].

Moreover, some group of pesticides consist of neonicotinoid pesticides such as guadipyr, imidacloprid, and thiacloprid. This group of pesticides are new and they constitutes the most utilized insecticides Goulson, [96] most especially for their non-target effect on beneficial microorganisms [97]. Conversely, some scientist have demonstrated a contrary opinion to these discovery [98–100], this include their influence on bees [101]. The effect of these pesticides has been shown on reproductive systems of animals and endocrine [102]. Also, the effect of neonicotinoids has been established to induce rise in the manifestation of the enzyme aromatase that could stimulate the development of the enzyme aromatase most especially during their developmental stages [103].

13.4 Conclusion and Recommendation to Knowledge

This chapter has provided detailed information on the application of pesticides in agriculture. The merits and demerits of pesticides were also highlighted. The modes of action through which most of these pesticides are applied were also stated. The detailed information on the following pesticides was also discussed such as organochlorine, organophosphorus, carbamate pesticides, and other classes of chemical pesticides. Therefore, there is a need to search for a sustainable and ecological friendly biopesticides that could mitigates all the health and environmental hazards associated with synthetic pesticides. The application of bio surfactants could reduce the adverse effect of these pesticides. There is a need for policy that will support the application of biopesticides as an alternative for synthetic pesticides. This will led to increase in food production as well as led to a cleaner environment. This will also led to drastic reduction and mitigate the high level of pesticides residue on agricultural products.

References

1. World Bank, *World Development Report*, Agriculture for Development; World Bank, Washington, DC, USA, 2008.
2. Donatelli, M., Magarey, R.D., Bregaglio, S., Willocquet, L., Whish, J.P., Savary, S., Modelling the impacts of pests and diseases on agricultural systems. *Agric. Syst.*, 155, 213–224, 2017.
3. *International Code of Conduct on Pesticide Management. In:* Annotated list of Guidelines for the implementation of the International Code of Conduct on Pesticide Management, Knowledge Exchange, Research and Extension, FAO, Viale delle Terme di Caracalla, 00153 Rome, Italy 2014.
4. Aktar, M.W., Sengupta, D., Chowdhury, A., Impact of pesticides use in agriculture: Their benefits and hazards. *Interdiscip. Toxicol.*, 2, 1–12, 2009.
5. Igbedioh, S.O., Effects of Agricultural Pesticides on Humans, Animals, and Higher Plants in Developing Countries. *Arch. Environ. Health Int. J.*, 46, 218–224, 2010.
6. World Health Organization, *Public Health Impact of Pesticides Used in Agriculture*, World Health Organization, England, 1990.
7. Alewu, B. and Nosiri, C., Pesticides and human health, in: *Pesticides in the Modern World – Effects of Pesticides Exposure*, M. Stoytcheva (Ed.), pp. 231–50, InTech, University Campus STeP Ri Slavka Krautzeka 83/A 51000 Rijeka, Croatia 2011.
8. NSW EPA, What Are Pesticides and How Do They Work?, 2013, https://www.epa.nsw.gov.au/your-environment/pesticides/pesticide-use-nsw.
9. Hoffman, R.S., Capel, P.D., Larson., S.J., Comparison of pesticides in eight U.S. urban streams. *Environ. Toxicol. Chem.*, 19, 2249–2258, https//:doi:10.1002/ etc.5620190915, 2000.

10. Canadian Cancer Society, Cosmetic Pesticides. Information Brief, 2013, https://www.cancer.ca/~/media/cancer.ca/AB/get%20involved/take%20action/CosmeticPesticides-InformationBrief-AB.pdf.
11. Johnston, J.J., *Introduction to Pesticides and Wildlife*, p. 589, USDA National Wildlife Research Center–Staff Publications, Fort Collins, Colorado, USA, 2001, https://digitalcommons.unl.edu/icwdm_usdanwrc/589/.
12. Hayes, T.B., Case, P., Chui, S., Chung, D., Haeffele, C., Haston, K. *et al.*, Pesticide mixtures, endocrine disruption, and amphibian declines: are we underestimating the impact? *Environ. Health Perspect.*, 114, 40–50, https//:doi:10.1289/ehp.8051, 2006.
13. Sanborn, M., Kerr, K.J., Sanin, L.H., Cole, D.C., Bassil, K.L., Vakil, C., Non-cancer health effects of pesticides. Systematic review and implications for family doctors. *Can. Fam. Physician*, 53, 1712–20, 2007.
14. Pimentel, D. and Burgess, M., Environmental and economic costs of the application of pesticides primarily in the United States, in: *Integrated Pest Management*, D. Pimentel and R. Peshin (Eds.), pp. 47–71, Springer Science + Business Media Dordrecht, New York, Heidelberg, Dordrecht, London, 2014.
15. Mnif., W., Hassine, A.I.H., Bouaziz, A., Bartegi, A., Thomas, O., Roig, B., Effect of endocrine disruptor pesticides: a review. *Int. J. Environ. Res. Public Health*, 8, 2265–2203, https//:doi:10.3390/ijerph8062265, 2011.
16. Goulson, D., Ecology: pesticides linked to bird declines. *Nature*, 511, 295–6, https//:doi:10.1038/nature13642, 2014.
17. Zheng, S., Chen, B., Qiu, X., Chen, M., Ma, Z., Yu, X., Distribution and risk assessment of 82 pesticides in Jiulong River and estuary. *Chemosphere*, 144, 1177–1192, https//:doi:10.1016/j.chemosphere.2015.09.050, 2016.
18. Pirsaheb, M., Limoee, M., Namdari, F., Khamutian, R., Organochlorine pesticides residue in breast milk: a systematic review. *Med. J. Islam. Repub. Iran*, 29, 228, 2015.
19. Sherwani, S.I., Arif, I.A., Khan, H.A., Modes of Action of Different Classes of Herbicides, Herbicides, in: *Physiology of Action, and Safety,* Andrew Price, Jessica Kelton and Lina Sarunaite, IntechOpen, 2015, University Campus STeP Ri Slavka Krautzeka 83/A 51000 Rijeka, Croatia. htpps//: doi: 10.5772/61779.
20. Wesseling, C., McConnell, R., Partanen, T., Hogstedt, C., Agricultural pesticide use in developing countries: health effects and research needs. *Int. J. Health Serv.*, 27, 273–308, https//:doi:10.2190/E259-N3AH-TA1Y-H591, 1997.
21. Eskenazi, B., Bradman, A., Castorina, R., Exposures of children to organo-phos-phate pesticides and their potential adverse health effects. *Environ. Health Perspect.*, 107, 409–19, https//:doi:10.1289/ehp.99107s3409, 1999.
22. Jaeger, J.W., Carlson, I.H., Porter, W.P., Endocrine, immune, and behavioural effects of aldicarb (carbamate), atrazine (triazine) and nitrate (fertilizer) mixtures at groundwater concentrations. *Toxicol. Ind. Health*, 15, 133–51, https//:doi:10.1177/074823379901500111, 1999.

23. García, A.M., Pesticide exposure and women's health. *Am. J. Ind. Med.*, 44, 584–94, https//:doi:10.1002/ajim.10256, 2003.
24. Gunnell, D., Eddleston, M., Phillips, M.R., Konradsen, F., The global distribution of fatal pesticide self-poisoning: systematic review. *BMC Public Health*, 7, 357, https//:doi:10.1186/1471-2458-7-357, 2007.
25. McGill, A.E. and Robinson, J., Organochlorine insecticide residues in complete prepared meals: a 12-month survey in S.E. England. *Food Cosmet. Toxicol.*, 6, 45–57, https//:doi:10.1016/0015-6264(68)90080-1, 1968.
26. Cabras, P. and Angioni, A., Pesticide residues in grapes, wine, and their processing products. *J. Agric. Food Chem.*, 48, 967–73, https//:doi:10.1021/jf990727a, 2000.
27. Damgaard, I.N., Skakkebaek, N.E., Toppari, J., Virtanen, H.E., Shen, H., Schramm, K.W. *et al.*, Persistent pesticides in human breast milk and cryptorchidism. *Environ. Health Perspect.*, 114, 1133–1138, https/:doi:10.1289/ehp.8741, 2006.
28. Buscail, C., Chevrier, C., Serrano, T., Pelé, F., Monfort, C., Cordier, S. *et al.*, Prenatal pesticide exposure and otitis media during early childhood in the PELAGIE mother-child cohort. *Occup. Environ. Med.*, 72, 12, 837–44, https/:doi:10.1136/ oemed-2015-103039, 2015.
29. Lu, D., Wang, D., Ni, R., Lin, Y., Feng, C., Xu, Q. *et al.*, Organochlorine pesticides and their metabolites in human breast milk from Shanghai, China. *Environ. Sci. Pollut. Res. Int.*, 22, 9293–306, https//:doi:10.1007/s11356-015-4072-z, 2015.
30. Lichtenberg, E., Economics of Pesticide Use and Regulation, in: *Reference Module in Earth Systems and Environmental Sciences Encyclopedia of Energy, Natural Resource, and Environmental Economics*, vol. 3, pp. 86–97, 2013, https://doi.org/10.1016/B978-0-12-375067-9.00092-9.
31. Watanabe, H., Inao, K., Vu, S.H., Phong, T.K., Ishihara, S., Takagi, K., Tournebize, J., Pesticide Exposure Assessment in Rice Paddy Areas: A Japanese Perspective. In Capri, E & Karpouzas, D (Eds.) *Pesticide Risk Assessment in Rice Paddies: Theory and Pratice*, pp. 167–214, Elsevier, The Netherlands.
32. Cheremisin, N.P. and Rosenfeld, P., *Handbook of Pollution Prevention and Cleaner Production: Best Practices in the Agrochemical Industry*, 3, 1–24, Elsevier B.V. Science Direct 2011. https://doi.org/10.1016/B978-1-4377-7825-0.00001-7.
33. Liu, Y., Pan, X., Li, J., A 1961–2010 record of fertilizer use, pesticide application and cereal yields: a review. *Agron. Sustain. Dev., Springer Verlag/EDP Sciences/INRA*, 35, 1, 83–93, 2015.
34. Rosenfeld, P. E., and Feng, L. G. H. Pesticides. *Risks of Haz Wastes*, 127–154, 2011. https://doi.org/:10.1016/b978-1-4377-7842-7.00011-8.
35. Sanchez-Bayo, F., *Impacts of Agricultural Pesticides on Terrestrial Ecosystems*, pp. 63–87, Centre for Ecotoxicology, University of Technology Sydney, Australia. Bentham Science Publishers Ltd. *Ecol. Impacts Toxic Chem*, 63–87, 2011.

36. Pandya, I.Y., Pesticides and Their Applications in Agriculture. *Int. J.*, 2, 2, 894–900, 2018.
37. Maton, S.M., Dodo, J.D., Nesla, R.S., Ali, A.Y., Environmental Impact of Pesticides Usage on Farmlands in Nigeria. *Int. J Inno. Res. Dev.*, 5, 4, 311–317, 2016.
38. Mattah, M.M., Precious, A.D.M., Futag, G., Pesticide Application among Farmers in the Catchment of Ashaiman Irrigation Scheme of Ghana: Health Implications. *J. Environ. Med.*, 1–7, 2015. https://doi.org/10.1155/2015/547272, 2015.
39. Lushchak, V.I., Matviishyn, T.M., Husak, V.V., Storey, J.M., Storey, K.B., Pesticide toxicity: a mechanistic approach. *EXCLI J.*, 17, 1101–1136, 2018.
40. Rajashekar, Y. and Shivanandappa, T., Mode of Action of the Natural Insecticide, Decaleside Involves Sodium Pump Inhibition. *PLoS ONE*, 12, 1, e0170836, https://doi.org/10.1371/journal.pone.0170836, 2017.
41. Casida, J.E., Pest Toxicology: The Primary Mechanisms of Pesticide Action. *Chem. Res. Toxicol.*, 22, 4, 609–19, https//:doi: 10.1021/tx8004949, 2009.
42. Casida, J.E., Mode of Action of Pesticides, Metabolism of Organophosphorus Insecticides in Relation to Their Antiesterase Activity, Stability, and Residual Properties. *J. Agric. Food Chem.*, 4, 9, 772–785, https://doi.org/10.1021/jf60067a001, 1956.
43. United States Environmental Protection Agency, Insecticides, 2017, https://www.epa.gov/caddis-vol2/insecticides.
44. Turusov, V., Rakitsky, V., Tomatis, L., Dichlorodiphenyltrichloroethane (DDT): ubiquity, persistence, and risks. *Environ. Health Perspect.*, 110, 125–8, https//:doi:10.1289/ehp.02110125, 2002.
45. Van den Berg, H., Global status of DDT and its alternatives for use in vector control to prevent disease. *Environ. Health Perspect.*, 117, 1656–63, https://doi:10.1289/ehp.0900785, 2009.
46. Rodríguez-Alcalá, L.M., Sá, C., Pimentel, L.L., Pestana, D., Teixeira, D., Faria., A. *et al.*, Endocrine disruptor DDE associated with a high-fat diet enhances the impairment of liver fatty acid composition in rats. *J. Agric. Food Chem.*, 63, 9341–9348, https//:doi:10.1021/acs.jafc.5b03274, 2015.
47. Tiemann, U., *In vivo* and *in vitro* effects of the organochlorine pesticides DDT, TCPM, methoxychlor, and lindane on the female reproductive tract of mammals: a review. *Reprod. Toxicol.*, 25, 316–326, https://doi:10.1016/j.reprotox.2008.03.002, 2008.
48. Freire, C., Koifman, R.J., Koifman, S., Hematological and hepatic alterations in brazilian population heavily exposed to organochlorine pesticides. *J. Toxicol. Environ. Health A*, 78, 534–548, https//:doi:10.1080/15287394.2014.999396, 2015.
49. Lemaire, G., Terouanne, B., Mauvais, P., Michel, S., Rahman, R., Effect of organo-chlorine pesticides on human androgen receptor activation *in vitro*. *Toxicol. Appl. Pharmacol.*, 196, 235–246, https//:doi:10.1016/j.taap.2003.12.011, 2004.

50. Karami-Mohajeri, S. and Abdollahi, M., Toxic influence of organophosphate, carbamate, and organochlorine pesticides on cellular metabolism of lipids, proteins, and carbohydrates: A systematic review. *Hum. Exp. Toxicol.*, 30, 9, 1119–1110, https//:doi:10.1177/0960327110388959, 2011.
51. Calle, E.E., Frumkin, H., Henley, S.J., Savitz, D.A., Thun, M.J., Organochlorines and breast cancer risk. *CA Cancer J. Clin.*, 52, 301–309, https//:doi:10.3322/canjclin.52.5.301, 2002.
52. Robinson, T., Ali, U., Mahmood, A., Chaudhry, M.J., Li, J., Zhang, G. *et al.*, Concentrations and patterns of organochlorines (OCs) in various fish species from the Indus River, Pakistan: a human health risk assessment. *Sci. Total Environ.*, 541, 1232–1242, https//:doi:10.1016/j.scitotenv.2015.10.002, 2015.
53. Jaga, K. and Dharmani, C., Sources of exposure to and public health implications of organophosphate pesticides. *Rev. Panam. Salud Publ.*, 14, 171–185, https//:doi:10.1590/S1020-49892003000800004, 2003.
54. Li, Y.S., He, X., Ma, K., Wu, Y.P., Li, B.X., The effect of exposure to atrazine on dopaminergic development in pubertal male SD rats. *Birth Defects Res. B Dev. Reprod. Toxicol.*, 104, 184–189, https//:doi:10.1002/bdrb.21151, 2015.
55. Hung, D.Z., Yang, H.J., Li, Y.F., Lin, C.L., Chang, S.Y., Sung, F.V. *et al.*, The long-term effects of organophosphates poisoning as a risk factor of CVDs: a nationwide population-based cohort study. *PLoS One*, 10, e0137632, https//:doi:10.1371/ journal.pone.0137632, 2015.
56. Jamal, F., Haque, Q.S., Singh, S., Rastogi, S.K., The influence of organophosphate and carbamate on sperm chromatin and reproductive hormones among pesticide sprayers. *Toxicol. Ind. Health*, 32, 1527–1536, https//:doi:10.1177/0748233714568175, 2015.
57. Lin, J.N., Lin, C.L., Lin, M.C., Lai, C.H., Lin, H.H., Yang, C.H. *et al.*, Increased risk of dementia in patients with acute organophosphate and carbamate poisoning: a nationwide population-based cohort study. *Medicine (Baltimore)*, 94, e1187, https//:doi:10.1097/MD.0000000000001187, 2015.
58. Eskenazi, B., Harley, K., Bradman, A., Fenster, L., Wolff, M., Engel, S. *et al.*, In utero pesticide exposure and neurodevelopment in three NIEHS/environmental agency children's centre birth cohorts. *Epidemiology*, 17, S103, https//:doi:10.1097/00001648-200611001-00249, 2006.
59. Eskenazi, B., Harley, K., Bradman, A., Weltzien., E., Jewel, N.P., Barr., D.B. *et al.*, Association of *in Utero* organophosphate pesticide exposure and fetal growth and length of gestation in an agricultural population. *Environ. Health Perspect.*, 112, 1116–1124, https//:doi:10.1289/ehp.6789, 2004.
60. Rauh, V.A., Garcia, W.E., Whyatt, R.M., Horton, M.K., Barr, D.B., Louis, E.D., Prenatal exposure to the organophosphate pesticide chlorpyrifos and childhood tremor. *Neurotoxicology*, 51, 80–86, https//:doi:10.1016/j.neuro.2015.09.004, 2015.
61. Williams, G.M., Kroes, R., Munro, I.C., Safety evaluation and risk assessment of the herbicide Roundup and its active ingredient, glyphosate, for humans. *Regul. Toxicol. Pharm.*, 31, 117–1165, https//:doi:10.1006/rtph.1999.1371, 2000.

62. Mink, P.J., Mandel, J.S., Lundin, J.I., Sceurman, B.K., Epidemiologic studies of glyphosate and non-cancer health outcomes. *Regul. Toxicol. Pharm.*, 61, 172–184, https//:doi:10.1016/j.yrtph.2011.07.006, 2011.
63. Mink, P.J., Mandel, J.S., Sceurman, B.K., Lundin, J.I., Epidemiologic studies of glyphosate and cancer: a review. *Regul. Toxicol. Pharm.*, 63, 440–452, https//:doi:10.1016/j.yrtph.2012.05.012, 2012.
64. Cuhra, M., Traavik, T., Bøhn, T., Clone- and age-dependent toxicity of a glyphosate commercial formulation and its active ingredient in Daphnia magna. *Ecotoxicology*, 22, 251–262, https//:doi:10.1007/s10646-012-1021-1, 2013.
65. Campbell, A.W., Glyphosate: its effects on humans. *Altern. Ther. Health Med.*, 20, 9–11, 2014.
66. Samsel, A. and Seneff, S., Glyphosate's suppression of cytochrome P450 enzymes and amino acid biosynthesis by the gut microbiome: pathways to modern diseases. *Entropy*, 15, 1416–1463, https//:doi:10.3390/e15041416, 2013.
67. Woodburn, A.T., Glyphosate: production, pricing and use worldwide. *Pest Manag. Sci.*, 56, 309–312, https//:doi:10.1002/(SICI)1526-4998(200004), 2000.
68. Bonny, S., Genetically modified glyphosate-tolerant soybean in the USA: adoption factors, impacts and prospects. A review. *Agron. Sustain. Dev.*, 28, 21–32, https//:doi:10.1051/agro:2007044, 2008.
69. Duke, S.O. and Powles, S.B., Glyphosate-resistant crops and weeds: now and in the future. *AgBioForum*, 12, 346–357, 2009.
70. Thongprakaisang, S., Thiantanawat, A., Rangkadilok, N., Suriyo, T., Satayavivad, J., Glyphosate induces human breast cancer cells growth via estrogen receptors. *Food Chem. Toxicol.*, 59, 129–136, https//:doi:10.1016/j.fct.2013.05.057, 2013.
71. Fukutake, M., Takahashi, M., Ishida, K., Kawamura, H., Sugimura, T., Wakabayashi, K., Quantification of genistein and genistin in soybeans and soybean products. *Food Chem. Toxicol.*, 34, 457–461, https//:doi:10.1016/0278-6915(96)87355-8, 1996.
72. Swanson, N.L., Leu, A., Abrahamson, J., Wallet, B., Genetically engineered crops, glyphosate and the deterioration of health in the United States of America. *J. Org. Syst.*, 9, 6–37, 2014.
73. Kwiatkowska, M., Pawel., J., Bukowska, B., Glyphosate and its formulations – toxicity, occupational and environmental exposure (Article in Polish). *Med. Pr.*, 64, 717–729, https//:doi:10.13075/mp.5893.2013.0059, 2013.
74. George, J., Prasad, S., Mahmood, Z., Shukla, Y., Studies on glyphosate-induced carcinogenicity in mouse skin: a proteomic approach. *J. Proteomics*, 73, 951–964, https//:doi:10.1016/j.jprot.2009.12.008, 2010.
75. Samsel, A. and Seneff, S., Glyphosate, pathways to modern diseases III: manganese, neurological diseases, and associated pathologies. *Surg. Neurol. Int.*, 6, 45, https//:doi:10.4103/2152-7806.153876, 2015.
76. Pieniażek, D., Bukowska, B., Duda, W., Comparison of the effect of Roundup Ultra 360 SL pesticide and its active compound glyphosate on human

erythrocytes. *Pestic. Biochem. Physiol.*, 79, 58–63, https//:doi:10.1016/j.pestbp.2004.03.003, 2004.
77. de Liz Oliveira Cavalli, V.L., Cattani, D., Heinz Rieg, C.E., Pierozan, P., Zanatta, L., Benedetti Parisotto, E. *et al.*, Roundup disrupts male reproductive functions by triggering calcium-mediated cell death in rat testis and Sertoli cells. *Free Radical Biol. Med.*, 65, 335–346, https://doi:10.1016/j.freeradbiomed.2013.06.043, 2013.
78. Hedberg, D. and Wallin, M., Effects of Roundup and glyphosate formulations on intracellular transport, microtubules and actin filaments in *Xenopus laevis* melanophores. *Toxicol. in Vitro*, 24, 795–802, https//:doi:10.1016/j.tiv.2009.12.020, 2010.
79. Samsel, A. and Seneff, S., Glyphosate, pathways to modern diseases II: celiac sprue and gluten intolerance. *Interdiscip. Toxicol.*, 6, 159–184, https//:doi:10.2478/ intox-2013-0026, 2013.
80. Goad, E.R., Goad, J.T., Atieh, B.H., Gupta, R.C., Carbofuran-induced endocrine disruption in adult male rats. *Toxicol. Mech. Methods*, 14, 233–239, https//:doi:10.1080/15376520490434476, 2004.
81. Soloneski, S., Kujawski, M., Scuto, A., Larramendy, M.L., Carbamates: a study on genotoxic, cytotoxic, and apoptotic effects induced in Chinese hamster ovary (CHO-K1) cells. *Toxicol. in Vitro*, 29, 834–844, https//:doi:10.1016/j.tiv.2015.03.011, 2015.
82. Li, Q., Kobayashi, M., Kawada, T., Mechanism of ziram-induced apoptosis in human natural killer cells. *Int. J. Immunopathol. Pharmacol.*, 25, 883–891, https//:doi:10.1177/039463201202500406, 2012.
83. Li, Q., Kobayashi, M., Kawada, T., Carbamate pesticide-induced apoptosis and necrosis in human natural killer cells. *J. Biol. Regul. Homeost. Agents*, 28, 23–32, 2014.
84. Deninson, M.S., Phetan, D., Winter, G.M., Ziccardi, M.H., Cardaryl, a carbamate insecticide, is a ligand for the hepatic Ah (dioxin) receptor. *Toxicol. Appl. Pharmacol.*, 152, 406–414, 1998.
84. Li, Q., Kobayashi, M., Kawada, T., Carbamate pesticide-induced apoptosis in human T l ymphocytes. *Int. J. Environ. Res. Public Health*, 12, 3633–3645, 2015. https//:doi:10.3390/ijerph120403633.
85. Lifshitz, M., Shahak, E., Bolotin, A., Sofer, S., Carbamate poisoning in early childhood and in adults. *J. Toxicol. Clin. Toxicol.*, 5, 25–27, https://doi:10.3109/15563659709001161, 1997.
86. Zheng, T., Zahm, S.H., Cantor, K.P., Weisenburger, D.D., Zhang, Y., Blair, A., Agricultural exposure to carbamate pesticides and risk of non-hodgkin lymphoma. *J. Occup. Environ. Med.*, 43, 641–9, https//: doi:10.1097/00043764-200107000-00012, 2001.
87. Jin, Y., Wang, L., Chen, G., Lin, X., Miao, W., Fu, Z., Exposure of mice to atrazine and its metabolite diaminochlorotriazine elicits oxidative stress and endocrine disruption. *Environ. Toxicol. Pharmacol*, 37, 782–790, https//:doi:10.1016/j. etap.2014.02.014, 2014.

88. Kettles, M.K., Browning, S.R., Prince, T.S., Horstman, S.W., Triazine herbicide exposure and breast cancer incidence: an ecologic study of Kentucky counties. *Environ. Health Perspect.*, 105, 1222–1227, https//:doi:10.1289/ehp.971051222, 1997.
89. Song, Y., Jia, Z.C., Chen, J.Y., Hu, J.X., Zhang, L.S., Toxic effects of atrazine on reproductive system of male rats. *Biomed. Environ. Sci.*, 27, 281–288, https//:doi:10.3967/bes2014.050, 2014.
90. Breckenridge, C.B., Coder, P.S., Tisdel, M.O., Simpkins, J.W., Yi, K.D., Foradori, C.D. *et al.*, Effect of age, duration of exposure, and dose of atrazine on sexual maturation and the luteinizing hormone surge in the female Sprague-Dawley rat. *Birth Defects Res. B Dev. Reprod. Toxicol.*, 104, 204–217, https//:doi:10.1002/ bdrb.21154, 2015.
92. Kolaczinski, J.H. and Curtis, C.F., Chronic illness as a result of low-level exposure to synthetic pyrethroid insecticides: A review of the debate. *Food Chem. Toxicol.*, 42, 697–706, https//:doi:10.1016/j.fct.2003.12.008, 2004.
93. Ray, D.E. and Fry, J.R., A reassessment of the neurotoxicity of pyrethroid insecticides. *Pharmacol. Ther.*, 111, 174–193, https://doi.org/10.1016/j.pharmthera.2005.10.003, 2006.
94. Shafer, T.J., Meyer, D.A., Crofton, K.M., Developmental neurotoxicity of pyrethroid insecticides: critical review and future research needs. *Environ. Health Perspect.*, 113, 123–136, https//:doi:10.1289/ehp.7254, 2005.
95. Syed, F., John, P.J., Soni, I., Neurodevelopmental consequences of gestational and lactational exposure to pyrethroids in rats. *Environ. Toxicol.*, 31(12), 1761–1770, 2015. https//:doi:10.1002/tox.22178.
96. Goulson, D., An overview of the environmental risks posed by neonicotinoid insecticides. *J. Appl. Ecol.*, 50, 977–987, https//:doi:10.1111/1365-2664.12111, 2013.
97. Jeschke, P. and Nauen, R., Neonicotinoids-from zero to hero in insecticide chemistry. *Pest Manag. Sci.*, 64, 1084–1098, November 2008. *Pest Manag. Sci.*, 64, 11, 1084–1098, 2013. https//:doi:10.1002/ps.1631.
98. Qi, S., Wang, C., Chen, X., Qin, Z., Li, X., Wang, C., Toxicity assessments with Daphnia magna of Guadipyr, a new neonicotinoid insecticide and studies of its effect on cetylcholinesterase (AChE), glutathione S-transferase (GST), catalase (CAT) and chitobiase activities. *Ecotoxicol. Environ. Saf.*, 98, 339–44, 2013.
99. Put, K., Bollens, T., Wackers, F., Pekas, A., Non-target effects of commonly used plant protection products in roses on the predatory mite *Euseius gallicus* Kreiter & Tixier (*Acari: Phytoseidae*). *Pest Manag. Sci.*, 72, 7, 1373–1380, https//:doi:10.1002/ps.4162, 2016.
100. Williams, G.R., Troxler, A., Retschnig, G., Roth, K., Yañez, O., Shutler, D. *et al.*, Neonicotinoid pesticides severely affect honey bee queens. *Sci. Rep.*, 5, 14621, https//:doi:10.1038/srep14621, 2015.
101. Wright, G.A., Softley, S., Earnshaw, H., Low doses of neonicotinoid pesticides in food rewards impair short-term olfactory memory in foraging-age honeybees. *Sci. Rep.*, 5, 15322, 2015.

102. Hoshi, N., Hirano, T., Omotehara, T., Tokumoto, J., Umemura, Y., Mantani, Y., Insight into the mechanism of reproductive dysfunction caused by neonicotinoid pesticides. *Biol. Pharm. Bull.*, 37, 1439–1443, https//:doi:10.1248/bpb.b14-00359, 2014.
103. Caron-Beaudoin, É., Denison, M.S., Sanderson, J.T., Effects of neonicotinoids on promoter-specific expression and activity of aromatase (CYP19) in human adrenocortical carcinoma (H295R) and primary umbilical vein endothelial (HUVEC) cells. *Toxicol. Sci.*, 149, 1, 134–144, https//:doi:10.1093/toxsci/kfv220, 2016.

Index

Printed in the USA/Agawam, MA
September 16, 2026

910661.010